European Technical Coatings Congress (ETCC 2018)

Towards a Bright Future, New Coating Technologies

Amsterdam, The Netherlands
26 - 29 June 2018

ISBN: 978-1-5108-7959-1

Printed from e-media with permission by:

Curran Associates, Inc.
57 Morehouse Lane
Red Hook, NY 12571

Some format issues inherent in the e-media version may also appear in this print version.

Printed by Curran Associates, Inc. (2019)

Additional copies of this publication are available from:

Curran Associates, Inc.
57 Morehouse Lane
Red Hook, NY 12571 USA
Phone: 845-758-0400
Fax: 845-758-2633
Email: curran@proceedings.com
Web: www.proceedings.com

ETCC 2018
EUROPEAN TECHNICAL COATINGS CONGRESS

TOWARDS A BRIGHT FUTURE, NEW COATING TECHNOLOGIES

BOOK OF ABSTRACTS

26-29 JUNE 2018 | BEURS VAN BERLAGE
AMSTERDAM, THE NETHERLANDS

ETCC2018 IS
ORGANIZED IN CORPORATION WITH:

Nederlandse Vereniging van Verftechnici (NVVT)

The Dutch Association of Paint Technicians

FATIPEC

The Federation of Associations of Technicians for Industry of Paints in European Countries

OCCA

The Oil & Colour Chemists' Association

The Oil & Colour Chemists' Association

CoSi

COATINGS SCIENCE INTERNATIONAL

Coatings Science International

ACKNOWLEDGEMENTS

SILVER SPONSORS

ORONTEC the measurement and process experts

WAB

WORLÉE seit 1851

BRONZE SPONSORS

AkzoNobel

Applied Graphene Materials

BÜHLER

DOW

FERRO Where innovation delivers performance™

IMERYS Carbonates

OTHER SPONSORS

ASK CHEMICALS

ATP ENGINEERING

BASF We create chemistry

covestro

EVONIK POWER TO CREATE

Grolman 1851 International Distribution

PRA

synthomer

vito

vlci

SUPPORTING UNIVERSITIES

KTH VETENSKAP OCH KONST

TU/e

SUPPORTING MEDIA

coatings an Open Access Journal by MDPI

Journal of Coatings Technology and Research

SCIENTIFIC COMMITTEES

ETCC2018
Jaap Akkerman
Catarina Esteves
Pieter Geurink
Leo van der Ven

Fatipec European Scientific Committee
Asterios Bakolas
Pier Luigi Bonora
Thomas Brock
Peter Collins
Janos Hajas,
A. Eugene Indeikin
Alain Lemor
Abrahimi Morteza
David Pierre
Wolfgang Reisser
Heinz-Günther Schulte
Jaromir Snuparek
Leo van der Ven
Maria Zielecka

CoSi Scientific Committee
Wolfgang Bremser
Rolf van Benthem
Keimpe van den Berg
Stuart Croll
Catherine Dekerckheer
Catarina Esteves
Matthew Gebhard
Horst HintzeBruening
Mats Johansson
Sören Kiil
Raghavan Subasri
Bert de With
Limin Wu

FATIPEC

COATINGS SCIENCE INTERNATIONAL

CONTENTS

Summer school abstracts 07

Oral presentations *(ordered per session)* 11

Poster presentations 119

Authors index 167

SUMMERSCHOOL

Summer School 1: Functional Coatings

Workshop Leader: Catarina Esteves
Eindhoven University of Technology, Eindhoven, the Netherlands.

Coatings have been used since the prehistory for artistic and protective purposes. The advances of technology enabled improvements of the standard coatings functions, such as adhesion, corrosion protection, high gloss, scratch resistance, hydrophobicity, etc. With the great blossom of research and development in the last decades, and particularly with contributions from the fields of polymer chemistry and nanomaterials, coatings with novel functions became available. These are nowadays referred as Functional Coatings and deliver significant added value for commercialization.

A new function may derive from a tuned interaction at the air-coating or liquid-coating interface, a new bulk property or an improved coating-substrate interaction. Many types of functional coatings have been developed. Among the ones with the highest recognizable commercial interest are coatings with self-cleaning, self-healing, anti-fouling, anti-microbial, anti-icing, conductive, magnetic, photo- and thermochromic properties. Furthermore, it would be highly desirable that coatings play an "active" role, for instance by being able to respond or adapt to the changes in the environment or to the actions by the users. Responsive and dynamic coatings are currently the vital challenge in the field of functional coatings.

This workshop will provide an overview of the field functional coatings. The focus will be on discussing the relevant chemical and/or physical processes which endow a new property/function, with examples for several of the most commercially relevant functionalities, rather than trying to cover all possible functional systems. The understanding of the theories behind the design of functional coatings, and of the associated advantages and requirements, will provide the tools for applying it on new products and fine-tune their properties and final performance.

1. Functional Polymer Coatings: Principles, Methods, and Applications of Latex Film Formation: Processes and Properties. Edited by L. Wu and J. Baghdachi, Wiley, 2015, ISBN: 978-1-118-51070-4.

Summer school 2: Film Formation of Waterborne Coatings

Workshop Leader: Joseph L. Keddie
University of Surrey, Guildford, United Kingdom.

This lecture and workshop will provide a deeper understanding of latex film formation, which is the multi-step process by which a stable colloidal suspension is transformed into a cohesive coating.1 We will consider each of the three main stages in the process. First, the evaporation of water causes the particle concentration and suspension viscosity to rise. We will examine why water concentration gradients can often develop vertically (in the direction normal to the substrate) and laterally (in the plane of the coating) during the evaporation stage. Second, when soft particles are packed together in contact, they will undergo deformation to fill the available volume, under the action of a number of driving forces. We will explore the parameters that determine the mechanisms of particle deformation. Third, polymer molecules will diffuse across particle boundaries and entangle in neighbouring particles. We will review how this interdiffusion step leads to greater film cohesion and strength. As part of our discussion, we will consider other complicating factors, such as crosslinking and the presence of surfactants at interfaces.

The workshop will include some examples of simple calculations to explore the effects of parameters (such as the evaporation rate, temperature, particle size and polymer viscosity) on the stages of film formation. (Attendees should bring a calculator.) Our emphasis will be on understanding the relevant physical processes, rather than on the details of specific chemical systems. The workshop will provide a greater understanding of how to control the processes and to tailor the coating properties.

1. J.L. Keddie and A.F. Routh (2010) Fundamentals of Latex Film Formation: Processes and Properties, Springer.

Summer school 3: Life cycle analysis of coatings

Workshop Leader: Laura Golsteijn[1] & Wijnand Broer[2]
[1]PRé Sustainability, the Netherlands, [2]CREM, the Netherlands

Sustainability is becoming an integral part of every business and thus also of coatings. Currently, there are various sustainability topics that are arising and being embedded in a company's daily business, such as climate change, criticality, circular economy, natural capital and others. This summer school will provide an overview of these topics, their relation to the coatings industry and to Life Cycle Assessment (LCA). LCA is a methodology for assessing the potential environmental impacts of a product or service over its entire life cycle. LCA can measure the environmental performance of a coatings product, a coatings company and it can also help identifying improvement opportunities. During the summer school we will zoom in on the use of LCA by means of a coatings' case study. It will be an interactive session in which we'll be switching between presentations and exercises with group discussions. At the end, participants will have a good understanding of LCA, its challenges as well as its strengths.

PLENARY LECTURE 1

PL.01 Dynamic Functional Materials

Ben Feringa
University of Groningen, Groningen, The Netherlands

This plenary will focus on the chemistry of the Nobel Price lecture 2016 and will highlight the first results within the 'Advanced Research Center – Chemical Building Blocks Consortium' (ARC-CBBC) and its flagship program on Coatings & Functional Materials.

No abstract submitted

PLENARY LECTURE 2

PL.02 Stratification in films of binary colloidal mixtures: opportunities for designed coating surfaces

Joseph Keddie[1]
[1]Physics, University of Surrey, Guildford, United Kingdom

Abstract content: Waterborne colloidal films are used in a wide range of everyday applications, such as adhesives, inks, pharmaceutical coatings, agricultural treatments, paints and varnishes. For some applications, it is advantageous for a film's top surface to have a structure and properties that differ from the bulk. For instance, there have been reports that the segregation of hard particles at the surface of a waterborne polymer film increases the scratch resistance.1 In another example, gold nanoparticles have been stratified at the top of a polymer film to create photonic and plasmonic effects.2 In binary mixtures of particles of two different sizes, there are fundamental questions concerning the conditions under which there will be stratification of the particles by size (i.e. layering) and by which mechanisms.3 Within the past couple of years, our research group has discovered – via simulations and experiments4 – that when binary blends of waterborne particles are dried in a film, self-stratification occurs with a layer of small particles developing at the top interface with air (as shown in the figures below). The stratification is driven by an osmotic pressure, created by a concentration gradient of the small particles, acting on the large particles. In this lecture, I will present our experiments and simulations showing that the stratification mechanism is more favourable with higher size ratios, with a greater concentration of small particles, and in more dilute suspensions.5 We are able to tune the stratification by adjusting the particle size ratios and concentrations through pH triggered swelling of polymer-stabilised particles. A new diffusion model has recently been developed6 that provides a framework to understand the parameters that determine the stratification in the dilute limit. I will present new results that support the theory for dilute suspensions but show deviations from the theory for concentrated systems. 7 Our deeper understanding opens up new strategies to manufacture stratified coatings with targeted surface properties.

1. Nunes et al. (2014) Prog. Org. Coat. 77, 1523-1530.
2. A. Utgenannt et al. (2016) ACS Nano, 10, 2232-42.
3. A. F. Routh (2013) Rep. Prog. Phys., 76, 1.
4. A. Fortini et al. (2016) Phys. Rev. Lett., 116, 118301.
5. I. Martín Fabiani et al. (2016) ACS Appl. Mater. Interf., 8, 34755–61.
6. J. Zhou et al. (2017) Phys. Rev. Lett., 118, 108002.
7. D. Makepeace et al. (2017) Soft Matter, 13, 6969.

Keywords: colloid, film formation, nanoparticles, surface functionality

ORAL PRESENTATIONS

PARALLEL SESSION 01: Legislation and LCA

S.01-1 Legislation in the EU paint industry

Jan van der Meulen
CEPE, Brussel, Belgium

It is CEPE's mission to work on behalf of its members on legislations that are creating a level playing field and are proportionate in costs.

There is not always an holistic approach by the legislator to initiate new regulations.
Could Life Cycle Analysis be a help in this?

S.01-2 Improving air quality of waterborne coating systems by novel non-voc, apeo-free, non-silicon based high performance wetting agent with built in biodegradability

Giuseppe Lista
TS&D, DOW Chemical Iberica s.l., Tarragona, Spain

The paint and coatings industries are responding to regulatory and market demand with a broad range of offerings to minimize impacts to the environment and air quality, while delivering excellent performance. Dow has developed an ethoxylated alcohol wetting agent to address these growing needs for lower odor, reduced VOC content and higher performance in water-based coatings.

This presentation will include an overview of the most recent VOC regulations as well as the Company's strategy to design solutions which minimize emissions and improve performance. Specifically, the ethoxylated alcohol wetting agent will be compared to TMDD* and silicon based products, used as industry standards, in several water-based applications, including:

- Automotive coating CED
- Pigmented metal coating
- Non-pigmented metal coating
- Wood coating
- Pressure sensitive adhesives
- Laminate adhesives

Within this research, various properties were measured based on final coatings and adhesives applications. The formulated coatings were applied to different substrates, and after standard drying, the following properties were tested:

- Standard short-time corrosion, tested by measuring heat stability and water and salt resistance
- Chemical resistance of dried film
- Dynamic surface tension on various low energy surfaces substrates
- Initial low foam and after 5 minutes
- De-wetting improvement
- VOC ISO 11890-2

The impact of the novel ethoxylated alcohol wetting agent on film performance was evaluated. Evaluation of cured and aged samples shows outperforming properties rather than industry standards in heat stability, reduced dynamic surface tension, de-wetting, chemical and water resistance.

Results concluded the ethoxylated alcohol wetting agent allows water-based solutions to wet substrates as well as solvent-based solutions, especially in low energy surfaces substrates. This versatile wetting agent was also shown to improve foam control and resistance to corrosion, chemicals and water, while increasing coating line speeds and producing no VOC emissions.

* 2,4,7,9-tetramethyl-5-decyne-4,7-diol (TMDD)

Keywords: corrosion resistance, Dynamic surface tension, non voc, Waterborne coating

S.01-3 Drivers for innovation, sustainability and transparency in the supply chain

Joost Broeders
Innovation & Sustainability, Baril Coatings B.V.,'s-Hertogenbosch, the Netherlands

Baril Coatings develops, produces and sells coatings for various substrates like steel, non ferrous metals, concrete, wood and engineering plastics.
As a family owned company, we are convinced that all our activities should lead to lowering the overall impact on both environmental and social aspects.
Already for more than 15 years Corporate Social Responsibility thinking and acting is common practice at Baril.
'More with less' is our motto and it is the basis for our actions in innovation, process optimisation and raws selection, but above all in application (at customers) and life time expectancy (durability) of our coating systems.
We will present our successful drivers in innovation and sustainable acting, being:
R&D focus on minimising material use in combination with prolonged durability/life time expectancy, all backed by evidence based policy (independent testing and certification).
Also focus on incorporation of biobased raw materials, though not 100% biobased at the moment, as quality/performance and application requirements should at least be similar to, preferably outperform, conventional products.

The outcome of two major innovation projects will be presented to illustrate above drivers. One about our patented DualCure chemistry and a second one about our Biobased Eco-label certified productline.
With the patented DualCure technology, our customers are enabled to minimize material use (2 layer thin coat system instead of 3-4 layer thick coat system), speed of application (1 day application instead of 3-4 day application), no/minimal system damage during transport to and construction on the building site (no touch up required after construction).
With our Biobased Pura product line, the professional painter is enabled to lower his environmental impact. The biobased products are Eco-label certified, have A++ indoor emission class certification, etc. All without compromises to quality and application performance friendlines.

Baril strongly advocates transparancy in the supply chain. Despite our efforts to collect LCA/LCI data of our raws (suppliers) we haven't succeeded yet. To overcome this challenge, we supported the development of the Sustainability Hotspot Scan. Together with another SME, we partnered up with TNO and MVO NL (CSR the Netherlands). The tool helps companies determine the sustainability opportunities and risks in the value chain of chemical products without the need for much data. The hotspot scan is an initial step towards cooperation between the chain partners and greater transparency about the real costs (including environmental and social costs) in the chemical chain.
The outcome of the tool can be used as starting point to open up discussion with suppliers (and of course also customers) to diminish, or even eliminate, the negative impact/hotspots in the supply chain.

At Baril Coatings we are convinced that we need to join forces to eventually meet the Sustainable Development Goals as outlined by the United Nations.
Yet, we are also certain that we shouldn't just discuss about action; we really need to act. At Baril we say 'Groen is doen', so we act.

References:

-DualCure patents: NL1034986, US 8889798, EP 2238210, CA 2713534EU Eco-Label: NL/044/002

Keywords: hotspotscan, innovation, sustainability, transparency

S.02-1 Multifunctional polymer coating systems based on novel cationic hindered amine monomers

Ahmed Abdelrahman
Exigence Technologies, Winnipeg, Canada

The global smart coatings market will expand at 29.8% CAGR during the period between 2017 and 2025 to reach $9.82 billion [1]. The majority of such growth is attributed to the demand on anticorrosion, antimicrobial, architecture and self-cleaning coatings from down stream industries like construction, electronics, automotive and military. Here, we report the synthesis of multifunctional polymer coating systems that demonstrate antimicrobial, antifouling properties and potential for anti-corrosion applications. For example, the design and the synthesis of novel class of cationic hindered amine monomers (CHM's) is illustrated. The cationic properties of the CHM's are provided by the integration of quaternary ammonium (QAS) and/or quaternary phosphonium (QPS) moieties whose corrosion inhibition properties are well established [2]. Our novel CNM's also feature a hindered amine in the form of 2,2,6,6-tetramethylpiperidine derivatives known for their application as hindered amine light stabilizers (HALS) [3]. A CHM's with surfactant-like properties (surfmers) were employed in emulsion polymerization to synthesize acrylic latex that have the cationic hindered amine surface functionalities covalently bound to the polymer backbone. Films prepared from CHM-acrylic latex exhibits chlorine-modulated total bacterial kill, 5 log reduction, against drug resistant E. Coli in 5% PBS in 10 min. Such an antimicrobial activity is maintained even after challenging the films with a variety of organic loads including FBS and sheep blood demonstrating strong resistance to protein adhesion and consequently antifouling properties. In addition, these CHM-acrylic films reveal free radical scavenging, halide pick up and UV stabilizer properties, which make such films potential candidates as protective coating and anti-corrosion applications. To prove the versatility of our novel CHM's, epoxy-functionalized CHM was synthesised and employed in the preparation of CHM-epoxy resin using commercially available amines. Coated coupons with such CHM-epoxy exhibits chlorine-modulated total bacterial kill, 6 log reduction, against E. Coli in 5% FBS in 10.

References:

[1] Transparency Market Research report: Smart Coatings Market: Global Industry Analysis, Size, Share, Growth, Trends, and Forecast 2017 – 2025
[2] Corrosion Science, 2000, 42 (8), 1307-1326
[3] Macromolecules, 2010, 43 (10), 4573–4583

Keywords: *acrylic latex, epoxy and polyurethane coatings, anticorrosion antimicroial antifouling coatings, smart coatings*

S.02-2 UV stable electrocoats for corrosion protection and structural bonding

Ulrich Christ, Nadine Göttl, Rolf Nothhelfer-Richter
Coating Systems & Paint technology, Fraunhofer IPA, Stuttgart, Germany

The coating process in the industry segment of mobile machinery, e.g. agricultural machines, is usually done after welding the framework with a variety of parts. The current painting process consists of an anti-corrosive primer and a weather resistant colored top coat.

The presented study is part of a research project investigating a new concept combining a new sequence in construction with the big advantage of a sole one coat system. This innovative technology enables inversion of the classical process: At first, the smaller pieces of the framework of the machine is coated with a single layer electrocoat (via cathodic electrodeposition CED). In the following step the parts are connected by structural bonding on the coating layer – instead of welding. This smart production enables the construction of the machines in modules, keeps the volume of an electrocoating dipping bath at a minimum level, and replaces the custom painting procedure with much overspray by an economical and ecological coating technology.

To realize this new technology, the coating systems has to perform as effective anti-corrosive system, provide a sufficient outdoor weather resistance in different color shades and must have sufficient adhesion strength for bonding. Therefore, the following coating properties are explored

- Modification of the paint formulation and development of different color shades
- Testing the anti-corrosive properties via salt spray test, humidity tests, cyclic corrosion tests
- Evaluation of barrier properties with electrochemical impedance spectroscopy
- Adhesion and other mechanical tests
- Testing UV- and weather resistance
- Determination of the process window of the electrocoat systems
- Investigation of structural bonding on the coating system

The cathodic electrodeposition coat is based on a low VOC hybrid acrylic-epoxy self-crosslinking binder system. Different formulations are modified and tested in order to improve anti-corrosive properties, the adhesion on steel substrates, to improve the UV- and weather resistance and the pre-requisites for structural bonding of the coating surface.

First results show that the electrocoat in different color shades (dry film thickness: 40 µm) shows anti-corrosive properties up to 1000 h salt spray test with low undercreepage values. UV- and weather resistance were acceptable up to 2000 h CAM 180, but depends on the color shade.

Keywords: electrocoat, corrosion protection, adhesion, weather resistance, structural bonding

S.02-3 Increased corrosion protection and durability of silica-PMMA hybrid coatings modified with cerium and lithium

Andressa Trentin, Samarah Harb, Celso Santilli, Peter Hammer
Physical Chemistry of Materials, São Paulo State University, Araraquara, Brazil

Organic-inorganic hybrid coatings prepared by the sol-gel route constitute an environmentally compliant alternative to replace current methods based on chromate passivation for corrosion protection of metallic surfaces. Silica-PMMA hybrids were prepared from the radical polymerization of methyl methacrylate (MMA) with 3-methacryloxipropiltrimethoxisilane (MPTS) using the thermal initiator benzoyl peroxide (BPO), followed by hydrolysis and condensation of tetraethoxysilane (TEOS) catalyzed in acidic conditions. Corrosion inhibitors in form of lithium and cerium salts were added in different concentrations (100 - 2000 ppm and 100 - 5000 ppm, respectively) into the inorganic precursor and the final solution was used to deposit few microns thick coatings on carbon steel (A1020) and aluminum alloy (AA7075) by dip coating. The structural properties were investigated using X-ray photoelectron spectroscopy (XPS), small angle X-ray scattering (SAXS), thermogravimetry (TG), optical, and atomic force microscopies (AFM). SAXS measurements revealed that the homogeneous distribution of a monomodal and diluted set of silica nanoparticles in the polymer matrix provided excellent barrier properties of the coating. The anticorrosive performance, evaluated using electrochemical impedance spectroscopy (EIS) in saline environment (3.5% NaCl), revealed for intact samples a corrosion resistance of up to 50 GΩ cm2, a value about 6 orders of magnitude higher than that of bare substrates. The immersion time evolution of cerium and lithium containing coatings presented an increase of the corrosion resistance after local failure (pitting) as a consequence of the inhibitor induced self-healing activity, leading to a durability of more than 9 months. Scratched samples were investigated before and after exposure to saline solution using XPS and Energy-dispersive X-ray spectroscopy (EDX). XPS results confirmed the presence of lithium and cerium in the damaged region and EDX mapping revealed after immersion an increase of oxygenated species into the scratch track. Compared to not modified samples, the extension of coatings lifetime evidences the beneficial effect of active inhibitors on the anticorrosive performance based on two different mechanisms: protective layer formation by leaching of lithium ions and presence of insoluble cerium oxides and hydroxides in affected film regions [1, 2].

Acknowledgments
Funding agencies: FAPESP and CNPq

References:
[1] P. Visser, Y. Liu, H. Terryn, J.M.C. Mol. J. Coat. Technol. Res. 13, 557–566 (2016).
[2] S.V. Harb, F.C. dos Santos, B.L. Caetano, S.H. Pulcinelli, C.V. Santilli, P. Hammer, RSC Adv. 5, 15414–15424 (2015).

Keywords: *corrosion protection, nanostructure, self-healing, Silica-PMMA hybrids*

S.03-1 Effective development of emulsion polymers via predictive working and high throughput screening

Sander van Loon
VLCI, Amsterdam, the Netherlands

A powerful theory for effective surfactant selection and monomer replacement for emulsion polymerization has been made practical by High Throughput (HT) screening. A relatively new approach has been proven to be much more efficient in generating useful parameters to develop (micro-) emulsions. It is based on the HLD-NAC theory, where profound predictions can be made about the type of emulsion (o/w, w/o) and the efficiency of a surfactant for a defined emulsion system. HLD-NAC takes into account the monomer or polymer number (EACN), temperature, salinity (and co-solvent), based upon which a surfactant can be selected with the right parameter (Cc) to achieve the thermodynamic stable region of the emulsion (HLD = 0). From there, it is possible to develop emulsion polymers, allowing for the most effective and best choice of surfactants at the lowest amount, and reduced energy input. This applies to a broad range of monomers, polymers and surfactants, even for combinations and bio-based materials. Although HLD-NAC is a powerful tool to rationalize and speed up research, HT brings this efficient approach further: it is very valuable for the determination of HLD-NAC parameters (simple screening) and the development of final emulsion polymers (complex formulation), in an efficient and consistent manner. In this presentation, the use of HT screening for determining EACN of monomers and selecting suitable surfactants (combination) will be given, and a proof of concept for emulsion polymerization based upon HLD-NAC will be showcased. With the obtained emulsion polymer, a wall paint has been made and tested, and the results will be discussed too.

S.03-2 New developments in low bake matte powder coatings

Nicole Willems, Leendert Molhoek, Matthijs Bos
R&D, DSM Coating Resins BV, Zwolle, the Netherlands

The increasing demand for reducing carbon footprint (CFP) of coating systems is also related to the existing curing temperatures used for powder coatings. This sustainability improvement creates an increased demand for low temperature curing (LTC) powder coatings without adverse effect on the curing times which is equally important for lowering manufacturing costs. Additionally, LTC powder coatings can also be used at higher curing temperatures for a shorter time.

Using LTC powder coatings also allows a wider range of parts to be coated. In the case of parts of a similar type, for example with very thin or very thick substrates, the lower peak temperature needed for powder coatings of this kind makes it possible for the coating to keep its nice appearance.

Lower cure temperatures also open up new applications for the powder coating process. Metal substrates which, for functional reasons, cannot be exposed to high temperatures or complete assem¬blies which also have restrictions on their peak temperature can be coated using these LTC powder coatings.

In the last six years DSM Coating Resins has developed a platform for low bake technologies for the metal market, which allow curing at temperatures of 160oC while obtaining the desired properties for indoor and outdoor applications. The first generation LTC powders coatings are developed for the high gloss segment. As addition to these high gloss LTC powder systems new LTC powder systems are developed with low glosses. To complete the LTC portfolio dry blend matte systems are developed existing of two powder formulations, each based on a different polyester resin. The challenge to develop a matte LTC dry blend was to develop resins, which should cure fast enough and show a sufficiently big reactivity difference between each other to end up with a matte powder coating. The developed matte LTC systems show a really unique combination of coating properties like mechanical properties, stable low glosses and are easy to handle. In the lecture we will discuss several extensive studies to the basic properties of the 2 new developed dry blend matte systems: a General Industry version and an Architectural version. Important properties as reactivity (LTC), robustness of matting (temperature and layer thickness dependency), mechanical properties, outdoor durability (QUV, WOM as well as Florida) and many other coating properties have been studied and will be shared.

And last but not least, the latest developments in this field will be presented too, a superdurable dry blend matte system and a recent developed one shot matte binder will complete this interesting study.

Keywords: elastomeric, roofing, Waterborne coating, waterproofing

S.03-3 Cationic binders for insulating paints on wood

Florian Ninnemann
Worlée Chemie GmbH, Lauenburg/Elbe, Germany

Wood is a very complicated substrate with different and even seasonally fluctuating properties. Beside the inhomogeneous structure or the shrinkage, painters have to face the so called "bleeding" of wood ingredients. Bleeding occurs as a discoloration of the coating and is caused by water soluble tannins, which are transported to the surface by the evaporating water.
One possibility to avoid bleeding is the use of fixating metals salts, which often does not induce the desired effect in full or leads to side-effects. Another option that is more and more considered is the use of cationic acrylic binders, that indeed show excellent insulating properties in wood primers or decorative coatings. They also have great other stain-blocking properties e.g. against nicotine.
In this presentation, we will focus on the use of cationic acrylic binders. The difference between cationic solutions and dispersions will be discussed as well as the difference in insulating properties. We will also present our results with regards to using different types of wood. For example woods like oak with a higher content of tannins are more difficult to be insulated.
One common fear that is often heard from paint manufacturers concerns the use of cationic binders in a plant where until now only anionic systems have been processed. We will talk about what needs to be considered when working with cationic polymers.
The results show that cationic polymers are an interesting option for insulating different types of wood and could be included in any test that involves stain blocking, whether tannins, nicotine or other stains.

Keywords: cationic, stain blocking, waterborne acrylics, wood primers

S.04-1 Combination of photopolymerization and aza-michael addition for coatings with enhanced properties

Celine Croutxe-Barghorn, Matthieu Retailleau, Ahmad Ibrahim, Xavier Allonas
LPIM, University of Haute-Alsace, Mulhouse cedex, France

A click chemistry synthetic strategy based on aza-Michael addition and radical photopolymerization was implemented to synthetize a poly (amino ester)–poly(acrylate) network. The whole process combines advantageous properties (low shrinkage, tunable properties…) without facing the presence of unreactive nucleophile groups, resulting in a final polymer with enhanced homogeneity and thermomechanical properties compared to conventional photocured acrylate networks. The potential of this three-step approach was investigated for applications in the field of shape memory polymers (SMP). Resulting photopolymers exhibit excellent shape fixity and recovery within a narrow temperature range. They also demonstrate higher scratch resistance. Finally, through the shape memory effect afforded by this process, they present a healing behavior under thermal activation after macroscopic scratches.

References:

(1) Retailleau, M.; Ibrahim, A.; Croutxé-Barghorn, C. ;Allonas, X., WO 2017009127A1
(2) Retailleau, M.; Ibrahim, A.; Croutxé-Barghorn, C.; Allonas, X.; Ley, C ; Le Nouen, D. ACS Macro Lett. 2015; 1327-31.
(3) Retailleau, M.; Ibrahim, A.; Croutxé-Barghorn, C.; Allonas, X., RSC Adv. 2016, 6 (52), 47130-47133.

Keywords: aza-Michael addition, coatings, photopolymerization, scratch resistance

S04-2 Photoinitiators - key components enabling an efficient accesses to novel materials

Kurt Dietliker
Department of Chemistry and Applied Biosciences, Swiss Federal Institute of Technology Zurich (ETH-Z), Zurich, Switzerland

Radiation curing is a well-established technology with widespread uses in industry. Outstanding properties, such as control of polymerization in time and space, fast processes, high product quality, energy efficiency and a favorable ecological footprint allow for uses spanning from well-established applications such as protective coatings, printing inks, adhesives or electronic materials to recent developments in additive manufacturing and biomedical applications.
A key component for all applications is the photoinitiator which is important both for further improvements of existing technologies and the realization of new applications. This paper illustrates with some selected examples how recent developments in photoinitiators are drivers for novel applications.
The recent introduction of photolatent bases made the vast field of materials crosslinked under base catalysis accessible for radiation curing. In addition to enabling the curing of resins, the photogenerated base can also be used as catalyst for processes such as sol-gel reactions of inorganic materials or the in-situ reduction of graphene oxide. By combining the two processes, the controlled and efficient design of organic-inorganic hybrid materials or nanocomposites with outstanding features, such as improved mechanical properties or thermal and electrical conductivity, becomes possible.
The fast introduction of UV/Vis-LEDs as light sources in radiation curing is also a strong driving force for new developments. The very long lifetime of LED light sources, energy savings due to immediate switch on/switch off of the lamp, or the absence of heat-emitting IR radiation are obvious technical and application advantages, which allow for uses not feasible with conventional lamps. However, photoinitiators hitherto available were developed for broad-band emitting light sources and had to be adapted to provide the necessary sensitivity with narrow-band LED for enabling these new applications.
Water-borne formulations become very attractive as UV-LED curable inkjet inks with a very low viscosity and non-migration properties required for packaging applications. Biomedical uses, extending from pharmaceutical applications to 3D printing in tissue engineering, are based on photopolymerization in an aqueous environment. These developments set new requirements to the photoinitiator and stimulated an intensive search for photoinitiators that are both water-soluble and sensitive to the LED emission.
Suitable functional groups allow the linking of photoinitiators onto selected surfaces, such as glass or cellulose nanocrystals. Irradiation of these materials in a monomer solution results in a rapid and efficient surface modification under mild conditions, inducing features such as controlled wettability, improved laminate strengths in composite materials, and tunable repulsive or resorption capabilities required for antifouling surfaces or medical devices.

Keywords: hybrid materials, LED, photoinitiator, radiation curing

S.04-3 Low-VOC resins: a useful toolbox to counter current and upcoming legal application restrictions

Marco-Yann Heuer, Fabian Eichenberger, Andreas Stuettgen
AT Industrial Coatings EMEA, EVONIK Resource Efficiency GmbH, Essen, Germany

Worldwide the demand for more eco-friendly coatings is still increasing, After US and European Initiatives in the past for reducing the VOC emissions also China decided in 2017 to join this way. A toolbox of new binder technologies based on silicone-hybrid-resins as well as polyester based resins offer a unique combination of low viscosity and high durability. Next to these properties also improvements like enhanced adhesion promotion, higher weatherstability, excellent color stability are achieved by using these resins. It is possible to create high-solid and ultra-high-solid coatings materials Restrictions in Scandinavia about a limited usage of isocyanate curing products in coatings industry restrict the portfolio of suitable resins. A special silicone-hybrid-resin technology with a non-isocyanat-curing mechanism based on hydrolysation-condensation-reaction allows a durable polymeric network for protective coatings. In conjunction with aliphatic alpha-omega-trialkoxy-urethane resins it is also possible to create different degrees of hardnesses without losing any of the excellent properties like weather or chemical resistance. So more freedom for the formulator is given without any invest into new production or application equipment.
By using this special resin technology it is possible to achieve ready-to-apply coatings with 90% solid content.
These different technologies allow the fulfillment of common and future legal restrictions of different coaings materials.

Keywords: binder, Low VOC, Non-Isocyanate curing, resin

PARALLEL SESSION 05: Biobased, sustainable technologies

S.05-1 Biobased building blocks and resins for coatings

Jacco van Haveren
Wageningen University and Research, Wageningen, the Netherlands

No abstract submitted

S.05-2 Novel thermoset resins based on technical lignin fractions

Marcus Jawerth[1], Mats Johansson[1], Stefan Lundmark[2], Claudio Gioia[1], Martin Lawoko[1]
[1]Fibre and Polymer Technology, KTH Royal Institute of Technology, Stockholm, Sweden
[2]Perstorp AB Innovation, Perstorp, Sweden

Lignin is an aromatic and abundant biomolecule that has gained interest as a material constituent the last decades. Its aromatic structure provides a molecularly rigid constituent to be utilized in high T^g applications. Lignin is readily found in the waste stream of the pulp and paper industry and has long been considered as nothing more than an energy source through incineration[1]. New methods however allow for retrieval of lignin more readily than before. One such method is the Lignoboost process which is used to retrieve lignin out of the black liquor of the Kraft process[2]. When working with lignin it is important to know its origin in terms of plant species as well as the process used for extraction. The monomeric composition of the lignin is different between different feedstocks and depending on how harsh the conditions of the extraction procedure are the structure will be severely altered. One simplified way to view the lignin is as a mixture of aromatic molecules with functional groups. The most common functional groups found in processed lignin are phenolic, aliphatic, and carboxylic acid hydroxyls as well as aldehydes.[1]

In previous work Lignoboost Kraft lignin was solvent fractionated[3], chemically modified and used as a thermoset resin constituent for new lignin-based materials[4]. All solvents that are used are considered green to emphasise a sustainable process[5]. The ethanol soluble fraction of Lignoboost Kraft lignin has been selectively allylated towards phenols to high conversions and then thermally cured using thiol-ene chemistry[4].

Now, efforts are made to further develop the concept of lignin based thiol-ene thermosets. The present study show how a selective allylation of lignin phenols can be utilized on different fraction from the Kraft process to design both resin and thermoset performance. The effect of non-phenolic functionalities on the performance is furthermore elaborated, all of which will influence the properties of the final material.

References:

Duval, A.; Lawoko, M., A review on lignin-based polymeric, micro- and nano-structured materials. Reactive and Functional Polymers 2014, 85, 78-96, DOI http://dx.doi.org/10.1016/j.reactfunctpolym.2014.09.017.

Tomani, P., The LignoBoost Process. Cellulose Chemistry and Technology 2010, 44 (1-3), 53-58,

Duval, A.; Vilaplana, F.; Crestini, C.; Lawoko, M., Solvent screening for the fractionation of industrial kraft lignin. Holzforschung 2015, 70 (1), 11-20, DOI 10.1515/hf-2014-0346.

Jawerth, M.; Johansson, M.; Lundmark, S.; Gioia, C.; Lawoko, M., Renewable Thiol–Ene Thermosets Based on Refined and Selectively Allylated Industrial Lignin. ACS Sustainable Chemistry & Engineering 2017, DOI 10.1021/acssuschemeng.7b02822.

Prat, D.; Wells, A.; Hayler, J.; Sneddon, H.; McElroy, C. R.; Abou-Shehada, S.; Dunn, P. J., CHEM21 selection guide of classical- and less classical-solvents. Green Chemistry 2016, 18 (1), 288-296, DOI 10.1039/C5GC01008J.

Keywords: Lignin, Thermoset, Thol

S.05-3 Self-cured alkyd resin using non-drying avocado seed oil as a material of regenerative resource

Chigozie Uzoh
Chemical Engineering, Nnamdi Azikiwe University, Awka, Nigeria

Self-cured alkyd resin was synthesized from avocado seed oil (ASO) via a two way alcoholysis-polycondensation methods. The raw ASO was evaluated for its physical and chemical properties and further epoxidized, hydroxylated dehydrated to obtain a desaturated ASO. The fatty acids present in the raw ASO were evaluated with GC-MS while structural composition of the raw ASO, desaturated ASO and ASO based alkyd resin were evaluated with FTIR analytical technique. A surrogate kinetic model was used to approximate the acid functional group conversion in the alkyd reactor. FTIR and GC-MS results confirmed that the oil was predominantly saturated. The oil content of avocado seed is found to be 55% wt which justifies its availability as an industrial crop. The FTIR result further confirmed esterification reaction of the oil with acid anhydrides. The results of iodine values show that the unsaturation level of the oil increased significantly. The alkyd resin C produced met the requirements of standard drying test (ASTM 1640-D) in presence of nano-ZnO pigment at 32°C; exhibit excellent adhesion, abrasion and chemical properties. The kinetic study further discloses that conversion (82%), viscosity (269cP) and molecular weight average (4434g/mol) were within the desired value in the state variable space.

Keywords: avocado seed oil; non-drying oil; alkyd resin; surface coating

S.06-1 Assessment of water permeability of organic coatings by eis and alternative techniques

Tomas Prosek [1,2], Dominique Thierry[2]
[1]Technopark Kralupy, University of Chemistry and Technology Prague, Prague, Czech Republic,
[2]French Corrosion Institute, Brest, France

The principal function of an organic coating is to form a barrier between corrosive environment and metal. Although paint systems used in coil-coating industry are not able to provide full barrier effect, they do limit the access of ions, oxygen and water to the metal surface. The goal of this work was to compare techniques used for water uptake and water diffusion measurements in organic coatings in order to identify those applicable for coil-coated materials, to assess their limits and reproducibility and to verify if complementary information can be obtained by using several of them. Namely, gravimetry in humid air, gravimetry in liquid water, thermogravimetry, electrochemical impedance spectroscopy (EIS) and a cup method using blind hole samples have been compared. Attention was paid also to evaluation and interpretation of obtained data.

Sensitivity of the traditional gravimetry technique is on the borderline to be applicable for relatively thin coil-coating paint systems. Because of the same reason, kinetic data are difficult to obtain and the measurement is time consuming. However, it provides straightforward and easy to interpret results. Humid air is preferable to liquid water as a source of water.

The newly developed cup method using blind hole samples is a modification of a simple and established technique for industrial coil-coated products. Sample preparation requires experience but the measurement itself is simple and rapid. Disadvantage of the technique is the impossibility to distinguish between water transport through paint bulk and eventual micro-defects.

EIS is a rapid technique but because of problematic capacitance extraction from both single-frequency data and spectra fitting by equivalent circuits, ambiguous connection between water content and capacitance changes and troublesome fitting of capacitance vs. time curves, results of EIS measurements need to be used cautiously as there is a large space for misinterpretation. Whereas results for diffusion rate were rather consistent, water uptake (solubility) was obviously exaggerated.

Beside EIS, the tested techniques gave consistent material ranking. Discrepancies in absolute water permeability data can be explained partly by technical reasons and limitations of the techniques discussed above and partly by different underlying mechanisms of water transport and accumulation.

Financial support of Member Research Consortium (MRC) "Corrosion properties of coil-coated products" is appreciated.

Keywords: corrosion protection, EIS, organic coating, permeability

S.06-2 The active corrosion protection performance of the epoxy ester coatings containing sba-15 mesoporous silica nanoparticles

Mohammad Mahdavian[2], Misagh Amini[1], Reza Naderi[1]
[1]School of Metallurgy and Materials Engineering, University of Tehran,
[2]Surface Coating and Corrosion, Institute for Color Science and Technology, Tehran, Iran

Nanoinorganic materials can be used as reservoir of corrosion inhibitors to perform smart coatings as well as self-healing properties. In this work, SBA-15 mesoporous silica were synthesized and loaded with 2-mercaptobenzothiazol (MBT) as corrosion inhibitor. SBA-15 powders were analyzed by the nitrogen adsorption-desorption technique and the morphology were investigated by the scanning electron microscopy. The evaluations showed that SBA-15 mesoporous materials have a cylindrical ordered structure with high specific surface area of 609 m^2/g and large pore volume of 1.1 cm^3/g. Hence, SBA-15 nanorods are the appropriate option to load with corrosion inhibitors and use in anticorrosion coatings. The epoxy ester coatings containing SBA-15 mesoporous silica were applied on mild steel panels. The effect of nanocontainers on the electrochemical and corrosion behaviors of the coatings was investigated in a 3.5 wt. % sodium chloride solution, using electrochemical impedance spectroscopy. Results revealed that incorporation of MBT-loaded SBA-15 improved the corrosion resistance of the epoxy ester coatings. In addition, self-healing performance was observed at short immersion periods for the one containing MBT-loaded SBA-15.

Keywords: mesoporous silica, corrosion inhibitor, smart coatings, self-healing properties

S.06-3 Trends in emission and odor testing in vehicles

Clarissa Ty, Christian Schaller
Research and Development, Pfinder KG, Böblingen, Germany

The issues of emission and odor are an increasing part of national regulations. In general emission and odor is a prominent issue for all paint and coating applications even for industrial utilization and private end-user paints like facade paints or interior deco paints. This presentation provides an overview of current national regulations and describes the most important test procedures using the example of cavity wax. As the cavity wax prevents vehicles from corrosion and thus represents an important part of modern automobiles. Therefore considerations of emission and odor which can be emitted in the vehicle interior by the cavity wax are unavoidable. The use of screening methods is a useful tool to give a conclusion about the emission potential right from the development level. Potential sources of emission and odor in different cavity wax technologies and their impact on resource management are examined. For example, the use of higher refined materials might be necessary. Chemical crosslinking reactions which lead to unpleasant smelling and emitting side products need to be avoided. This might be conceivable through the application of new binder technologies. A standardization and harmonization of the test methods for the objective determination of odors and emissions in the vehicle interior are necessary steps for the comparability of measurement results in new developments.

Keywords: cavity wax, emission, odor, regulation

S.07-1 Waterborne acrylic polyurethane roof waterproofing coatings meeting European standards

Jouko Vyörykkä[1], Janah Szewczyk[2], Phil Imbesi[2], Cristina Miotto[1]
[1]Dow Europe GmbH, Horgen, Switzerland,
[2]The Dow Chemical Company, Collegeville, USA

EU Construction Products Regulation requires a construction product to have a CE marking if the product is placed on the market in the European Economic Area and it is covered either by a harmonized European product standard or a European Technical Assessment. In the liquid applied roof waterproofing coating product category, European Technical Approval Guideline (ETAG) 005 can be used to achieve CE marking. This presentation focuses on technical requirements and tests associated with ETAG 005 when assessing durability of the liquid applied roofing membranes or full roofing assemblies in roof waterproofing applications. In Europe, the dominant chemistry in the liquid applied roof waterproofing market is polyurethane-based, and often it is delivered as a solvent-borne system. In this presentation, we introduce a newly developed waterborne acrylic polyurethane polymer for roof waterproofing applications. The new acrylic polyurethane polymer was developed to provide superior mechanical and water resistance properties in comparison to current flexible acrylic waterborne coating, while effectively matching the solvent borne polyurethane mechanical properties. By using hard and soft segments, it was possible to achieve superior mechanical properties in a broad temperature range, which is necessary for exterior roof waterproofing coatings that are exposed to a spectrum of weather conditions. The new polymer also has a better health and safety profile than solvent-borne polyurethanes used in the market. The waterproofing membranes formulated with the new waterborne acrylic polyurethane polymer has passed the ETAG 005 assessment demonstrating its capability to perform as a liquid applied roof waterproofing product in the European market and thus providing more sustainable and safe alternative to solvent borne polyurethanes.

Keywords: low bake, matte, powder coatings

S.07-2 New siloxane binders for high temperature coating applications

Edward Crowston, Michel Hrebicik
Technical Service and Development, Dow Perfromance Silicones, Seneffe, Belgium

The coatings world needs to offer (thicker) systems able to withstand higher temperatures, leading to better corrosion resistance, or protection from heat or fire. Organic binders show their thermal limitations. To fill the unmet needs, a new range of siloxanes was recently developed: room temperature film forming, waterborne or solventless binders. Through specific examples of applications, we'll review these different families of Si-based products: chemical reactivity, possible thicknesses, high elasticity... different end-properties will be further described, including Silicone chemistry, that will be briefly commented and orientated to understand how film and performances can be generated. This paint approach opens the door to easy insulation, softer (application) process, and broader protection to any metallic substrate.

Keywords: binder, high temperature, siloxane

S.07-3 Versatile tin free organometallic catalysts with good hydrolytic stability for polyurethane and organosilane applications

Marcel Groenewoud[1], Ravi Ravichandran[2], John Florio[2]
[1]Sales and Technique, King Industries, Amsterdam, the Netherlands,
[2]King Industries Inc., Norwalk, USA

The need for efficient, hydrolytically stable metal catalysts has never been greater in the polyurethane industry. Toxicity concerns have put regulatory pressure on the use of organotin. In addition, metal catalysts that are free from 2-EHA (2-ethylhexanoic acid), are desirable, as 2-EHA and its salts are classified as CMR substances category 2, as this classification leads to undesired labelling issues. Several novel organometallic catalysts have been developed as effective tin alternates for accelerating the polyol/polyisocyanate reactions. These new versatile catalysts are effective accelerators for a variety of polyurethane applications, including coatings, elastomers, sealants, and foams. In addition these catalysts are also effective in promoting ambient cure in both one and two component organosilane based polymers, utilized in adhesives, sealants and coatings as either coupling agents or crosslinkers. These new catalysts contain no 2-ethylhexanoic acid in the form of a free acid or as the metal carboxylate.

S.08-1 Solid coatings with triggered motility

Dirk J Broer
Institute for Complex Molecular Systems, Eindhoven University of Technology, Eindhoven, the Netherlands

Polymers that perform a programmed and reversible shape change have a wide application potential varying from micro-robotics to spacecrafts. A well-studied principle is based on a triggered and controlled change of the molecular order in liquid crystal polymer systems. We are utilizing similar techniques to change the topography of surfaces. This can be in the form of responsive cilia integrated in surfaces which can be utilized for transport of liquids or species. But it can also be a triggered change of the surface topography that affects the touch sensation and is of interest for haptic applications. As with the programmed shape changes, the underlying mechanism is a change of order parameter of the molecular ordered liquid crystal network coating. In addition to changes in molecular direction, volume is created originating from vibrating molecular entities in the polymer network which create dynamic molecular voids. This dynamic free volume leads to a temporary density decrease and the correlated volume increase which can be patterned and localized by an accurate control over the liquid crystal director. The triggers to actuate the coatings are temperature, light, pH, changes in environment or electrical fields. Bringing motion into coatings will open the route to completely new applications where coating actively will perform work or respond to changes in their environment. As example, a coating configuration will be shown that actively remove soil from its surface. In addition, the coatings can be modified to secrete liquids upon request, for instance to lubricate, protect or clean. Also here an example will be shown of a coating that emits liquid, either by light or by an electrical field, that changes the adhesive properties of species applied on its surface.

Keywords: dynamic topographies, liquid crystal polymer network, smart surfaces

S.08-2 Light from within: illuminating complex dynamics in coatings from the inside out

Joris Sprakel

No abstract submitted

S.08-3 Nanoscale chemical imaging of water-based urethane-acrylic hybrid coatings by photo-induced force microscopy

Qi Chen[1], Saskia van der Slot[1], Sven Kreisig[2], Mingwen Tian[2], Ron Peters[1], Joachim Loos[2]
[1]DSM Coating Resins, Waalwijk, the Netherlands,
[2]DSM Materials Science Center, Geleen, the Netherlands

Water-based coatings are made of dispersed polymer particles with different chemical species in the order of 50 to 500 nm. It is widely accepted that the distribution of chemical species at nanoscale would have great influence on the coating properties. For example, Urethane-Acrylic (U-A) hybrids are composite polymers that combine the properties of the two polymers. The advantages of bringing both materials in each individual particle, rather than mixing two separate dispersions, are well documented in the literature. The main hypothesis of the superior properties of hybrids over blends is that the two materials are more inter-mixed at a length scale below 50 nm, i.e. nano-phase separation occurs within the hybrids rather than the micro-phase separation that is usually found in blends.
The ability to identify and localize different chemical species at nanoscale has been recognized as one of the key challenges in coating research and materials science. Conventional techniques used to identify and localize different chemical species, such as Raman microscopy or FT-IR spectral imaging, can provide chemical identification but are diffraction limited in spatial resolution.
One emerging class of chemical imaging techniques that has the potential of achieving nanoscale spactial resolution are the AFM probe enhanced nanospectroscopy techniques. These include tip enhanced Raman spectroscopy, photo-thermal induced resonance and scattering scanning near-field optical microscopy. Recently, another technique within this class, infrared photo-induced force microscopy (PiFM), is made commercially available. PiFM detects photo-induced molecular polarizability of feature sizes down to the molecular level by mechanical detection of the force gradient of the interaction between the optically driven molecular dipole and its mirror image in a metal coated AFM tip. Applying this technique, it was possible to identify the components from U-A blends using lasers with wavelengths corresponding to the respective characteristic peaks of the individual materials. The high spatial resolution, which is down to 10-20 nm in X, Y and Z direction, in combination with a high sensitivity of the setup are demonstrated. By combining four quantum cascade lasers an IR like local spectrum can be constructed and spectral imaging can be realized. Further, the morphology of U-A hybrids is investigated using PiFM and it is confirmed that the phase separated domains of individual components are of a much lower size compared to those domains in blends. The obtained morphology is well expected when the different composition of the materials and the synthesis procedure are taken into consideration.
In conclusion, we will show in this presentation that PiFM has a clear potential in chemical imaging beyond diffraction limit on coating systems. This technique helps to better understand the relationship between phase separated morphology and coating properties, which will further enable coating scientists to design polymer architectures and particle structures that can influence the morphology and properties of the coatings.

Keywords: AFM, nanoscale chemical imaging, urethane-acrylic hybrids, water-based coatings

S.09-1 Optimal prediction of pks: rso modified alkyd resin polycondensation process using discrete-delayed observations, ann and rsm-ga techniques

Chigozie Uzoh[1], Ukeme Timothy[2], Faith Okey-Onyesolu [1]
[1]Chemical Engineering, Nnamdi Azikiwe University, Awka, Nigeria
[2]Polymer and Textile Engineering, Nnamdi Azikiwe University, Awka, Nigeria

Highlights

- RSM-GA and ANN were used to optimize a polycondensation process for the first time.
- Three output (responses) and five input variables were considered.
- RSM-GA performance were much better than ANN in all the responses.
- Non-dominated optimal route to achieve high quality resin economically were established

Alkyd resins are widely used in paint industry and although they have a long history about 70-100 years, today the developments in alkyds are still welcomed and innovations are still needed. Artificial neural network (ANN) and response surface methodology based on a 2^{5-1} fractional factorial design were used as tools for simulation and optimization of the polycondensation process for auto-oxidative drying alkyd resin from palm kernel stearin (PKS): rubber seed oil (RSO) blend of 70:30 ratio. A feed forward neural network model with Levenberg-Marquardt back propagation training algorithm was adapted to predict the responses (conversion Y_1, viscosity Y_2, and molecular weight average Y_3). The studied input variables were reaction time, temperature, catalyst concentration, oil ratio and stirring rate. The performance of the RSM and ANN model showed adequate prediction of the responses in terms of the process factors, with MRPD of ±4.47% (Y_1), ±2.08 %(Y_2), ±8.92% (Y_3) and ±6.50% (Y_1), ±3.31% (Y_2), ±10.20% (Y_3), respectively. The sensitivity analysis showed that, while reaction time is the most effective process parameter, the interaction of the five process variables produced the most significant effect on the studied responses with the overall minimum MSE of 0.079. The optimization task performed using genetic algorithm linked to the RSM model gave a viable, non-dominated optimal responses and optimum operating conditions regarding the route to high quality resin at reduced material and operational costs. Overall, coupled RSM-GA was found to be a better tool for modelling and optimization of the alkyd resin production.

Keywords: multivariate system, genetic algorithm, artificial neural network, alkyd resin

S.09-2 The future of carbon black pigments is green

Ad van Oorschot, Rik Leunissen
Black Bear Carbon B.V., Black Bear Carbon B.V., Nederweert, the Netherlands

Will address how Black Bear's green carbon black pigments are highly attractive technical substitutes for some conventional furnace and gas carbon blacks used in coatings:

- Introducing three new innovative "green" tire grades of carbon black – NEPtune C100, C80 & C60
- Performance in terms of colour strength and hyper-dispersibility in both water- and solvent-borne systems
- How these grades will drastically reduce the carbon footprint of coatings and contribute to preserving our planet

Keywords: carbon black, cradle to cradle certified, sustainable

S.09-3 Biorizon scaling up functionalized biobased aromatics for increased performance

Monique Wekking [2], Joop Groen[1]
[1]Biorizon, Bergen op Zoom, the Netherlands, [2]TNO, Den Haag, the Netherlands

Today, the majority of the commercially available biobased molecules is limited in their functionality to the groups of aliphatic acids and alcohols.1 This leaves the market for bio-based aromatics largely uncovered, while 40% of the bulk chemicals are based on aromatics.

When competing with naphtha-based aromatics, it is crucial that bio-based alternatives have at least similar performance in applications at a similar price; a premium for bio-based products is non-existent in the commodity market. For this reason, atom efficiency in the conversion of biomass into bio-aromatics is crucial for a viable business case. Therefore, to fully benefit from its economic potential, biomass, which is generally rich in oxygen functionalities, should be applied for the production of functionalized aromatics, e.g. phthalic acid. Vice versa, biomass cannot compete with naphtha in the production of BTX.

The number of routes towards functionalized bio-aromatics is limited. With respect to atom efficiency, biochemical conversion of biomass to aromatics could be promising, but these routes are generally limited in their requirement for relatively homogeneous and therefore costly feedstocks and inhibited by the toxicity of the aromatic products.

In search of higher yields and atom efficiencies, a biobased route towards methyl-phthalic acid was established. It's successful production at multi-kg-scale allowed tests by application partners, which revealed improved performance in coating and lubricant applications as compared to its petrochemical counterpart. Encouraged by the results obtained with methyl-phthalic acid, a consortium of industrial partners is currently establishing the multi-kg-scale production of hemi-mellitic acid (200 kton/a market), which has the promise of a significantly improved atom efficiency.

Now, 5 years after the start of the Biorizon initiative, a selection of routes, starting with the conversion of organic waste to furanics and their subsequent conversion to bio-aromatics, is demonstrated at lab-scale and scaled-up in bench-scale reactor skids (TRL 5). In the biphasic reactor-skid carbohydrate-rich feedstocks are converted into furanics, while in the BioHarT-skid furanics are subsequently converted into bio-aromatics, both at kg/h-scale. These skids are the first steps in Biorizon's mission is to enable commercial production of bio-aromatics by 2025.

Keywords: aromatics, biobased, performance, profitable

S.10-1 Monitoring corrosion and corrosion protection in bare and coated aluminium by optical-electrochemical hyphenated tools

Paul J. Denissen, Santiago J. Garcia
Novel Aerospace Materials, Aerospace Engineering Faculty, Delft University of Technology, Delft, the Netherlands

Electrochemical Impedance spectroscopy (EIS) is proven to be a useful technique to monitor corrosion processes and evaluate the properties of polymer coatings during the exposure to corrosive environments. Despite the ever-increasing literature on EIS analysis, data interpretation is still a major limitation as it requires a high level of user-experience and skills [1]. In this work we propose a new approach that reduces subjectivity and increases reliability of electrochemical signals interpretation (e.g. EIS) and their relation to the responsible macroscopic (surface) phenomena. For this, we hyphenated inexpensive ordinary optical microscopes to highly used electrochemical setups. The optical microscopy images are taken continuously during EIS measurements with the use of a specially designed opto-electrochemical setup. The optical images are acquired with a USB microscope and processed using a bimodal threshold algorithm. The software is then used to obtain characteristic parameters related to the evolution of on-going processes such as pitting corrosion, and in the case of coatings, delamination and underfilm corrosion-related phenomena. Besides this, the time-evolution and variation of the Open Circuit Potential (OCP) and low frequency impedance were used to monitor electrochemical processes. The setup and approach was validated by monitoring the corrosion behaviour of AA2024 in sodium chloride solutions. In this process it was possible to identify similar phenomena previously reported in time-consuming electrochemical [2] and post-mortem optical inspection [3]. The setup has also been used to evaluate the inhibiting efficiency of promising chromate replacing corrosion inhibitors for AA2024, i.e. Cerium Nitrate, Lithium Carbonate, Sodium diethyldithiocarbamate trihydrate, 2,5-dimercapto-1,3,4-thiadiazolate, 2-mercaptobenzothiazole, and 8-hydroxyquinoline. Furthermore, the setup has been proven as very effective to evaluate newly developed anticorrosive coatings using nanonetworks and algae exoskeleton carriers [4,5]. The hyphenation with in-situ optical analysis resulted in a better understanding of the corrosion inhibition and degradation processes revealing, in some cases, new information that could not be obtained by electrochemistry alone. Although the used data-treatment is currently still time-consuming, further progress on automated user-friendly software should make the live-interpretation of opto-electrochemical results possible for a broader audience outside the scientific community.

References:

[1] W. Tait, K. Handrich, S. Tait, J. Martin, in:, J. Scully, D. Silverman, M. Kendig (Eds.), Electrochem. Impedance Anal. Interpret., ASTM International, West Conshohocken (1993) 428–437.
[2] M.L. Zheludkevich, R. Serra, M.F. Montemor, K.A. Yasakau, I.M.M. Salvado, M.G.S. Ferreira, Electrochim. Acta 51 (2005) 208–217.
[3] A. Boag, A.E. Hughes, A.M. Glenn, T.H. Muster, D. McCulloch, Corros. Sci. 53 (2011) 17–26.
[4] P.J. Denissen, S.J. Garcia, Cerium-loaded algae exoskeletons for active corrosion protection of coated AA2024-T3, Corrosion Science, 128 (2017) 164-175.
[5] C.D. Dieleman, P.J. Denissen, S.J. Garcia, Inhibiting Nano-networks for Long-term Active Corrosion Protection of Damaged Coated Metals, to be submitted.

Keywords: AA2024, corosion inhibitor, EIS, optical microscopy

S.10-2 Anticorrosive primers with reduced zinc content - effect of zinc pigment modification

Matthias Wanner, Lukas Aktas, Michael Hilt
Coating Systems and Painting Technology, Fraunhofer IPA, Stuttgart, Germany

Zinc rich primers typically provide sound corrosion protection properties. High zinc pigment concentrations, however, are ecologically relevant and due to a lack of binder such high pigment concentrations may cause cohesive and adhesive weak points. The latter may become fatal, where insufficient surface preparation occurs. Since in practice imperfect or incomplete surface preparation or contamination after surface preparation happens much more frequently than presumed, measures which contribute to improved adhesion could help to counterbalance technical failures significantly. Such measures could support paint manufacturers, reduce customer complaints and partly relieve job coaters.
Therefore the following studies intend to explore, how far a combination of pigment variations with standard zinc pigment could contribute to improved anticorrosive properties, to decreased overall zinc content and therefore to enhanced adhesion on imperfectly prepared surfaces. Variations performed here comprise:

- shape, dimension and concentration of zinc pigment
- surface treatment of zinc pigment
- use of zinc alloy pigments
- metal precipitation on zinc pigment surface
- addition of inert conductivity additives.

Formulations based on these variations are applied, studied and evaluated regarding standardized corrosion lab tests, maritime outdoor weathering, basic properties in liquid and applied state and electrochemical test methods. First results of the latter two are already available. Regarding electrochemical characterizations, at least two new indication methods were found to be relevant in pilot tests performed:

1. Determination of the open circuit potential (OCP) of steel substrates furnished with the zinc primers to be characterized during a continued thermocyclic electrolytic loading. It was found that the loading duration necessary to increase the OCP above -0.7 V vs. calomel, which represents the transition to barrier phase in the "dual mode" theory about the mechanism of zinc primer, correlates with the performance of the respective primer in maritime splash water zone exposure.
2. The detection of the reversibility of water uptake – monitored during a hydrothermal cycling of the zinc primers applied and electrochemically detected via the coating capacitance – could deliver valuable information about the porosity and thermocyclically induced fatigue of the primer matrix and the extent of the formation of zinc oxidation products.

In addition, polarization experiments performed with thin layers of the zinc primer modifications mentioned above allowed an estimation of the anticorrosive effect of these modifications in comparison to standard zinc rich primers.

Keywords: characterization, corrosion protection, formulation, zinc primer

S.10-3 Investigation on a new durable surface treatment for titanium alloys to improve the adhesion properties for aerospace applications

Tobias Mertens
Surface Technology, Airbus, Bremen, Germany

In aerospace industry new design concepts offering weight savings are high in demand. Due to the increasing amount of composite materials also the percentage of titanium materials rises to ensure galvanic compatibility. However, it is known in aerospace industry that titanium materials exhibit issues in regard to long-term stable and durable adhesion. This paper evaluates presents a new anodizing treatment (N4-process) which leads to long-term stable and durable paint adhesion on titanium materials.
Within this study a new alkaline anodizing process (N4-Process) is used as a pre-treatment of the titanium alloy Ti6Al4V. For reference, an acidic etching process (HF/HNO3 etching solution) and a second anodizing procedure (Sulphuric Acid Anodizing, SAA) are employed to evaluate the nanostructure and adhesion properties obtained by the N4-Process. The influence of nano- and microstructure on adhesion properties was investigated using cross hatch test as well as a pull-of test. Additionally, the morphology of the titanium oxide layers and the fracture surfaces of the titanium oxide layers were analyzed via scanning electron microscopy (SEM) and transmission electron microscopy (TEM). Moreover the durability of the oxide layers after anodizing was investigated in hot-wet conditions. One aim of the work is to establish a better understanding of the long term stability of paints on titanium substrates. Within this work the correlation between the stability of the oxide layer and the paint adhesion were studied.

The SEM and TEM investigations show that the oxide layer after N4-process is nanostructured and consist out TiO_2 in rutile modification while the oxide layer after SAA is smooth and amorphous. The exposure of the both oxide layer to hot-wet conditions show significant differences. While the oxide layer after N4-process remains stable a crystallization effect of the SAA oxide was noticed.
The mechanical tests show that the area enlargement due to the anodizing processes enhances the adhesion of organic coatings also under hot-wet conditions. The results obtained, confirm the correlation between a nanostructured morphology and the durability of paints. While the SAA process leads to low adhesion properties in pull-off and cross hatch test the nanostructured surfaces after N4-process exhibits good long-term stable adhesion properties. May be the observed instability of the SAA oxide layer could also have an influence on the adhesion properties. Both the morphology and stability of titanium oxide layers are key to realize a long-term stable adhesion.

Keywords: adhesion, durability, surface treatment, titanium oxide layer

S.11-1 Novel isocyanate-free 1k moisture cure protective coatings based on vinyl neo esters and alkoxysilanes.

Denis Heymans[1], Catherine Romanowska[1], Bo Ngiabprasert[2], Marcelo Herszenhaut[3]
[1]Versatics, Hexion, Ottignies LLN, Belgium,
[2]Versatics, Hexion, Rotterdam, the Netherlands,
[3]Versatics, Hexion, Stafford, USA

Reactions between silanes and organic monomers offer an infinity of possibilities to optimize the performance of resins intended for coating applications. These copolymers are well-suited for high-durability protective applications. In particular, coatings based on acrylic-alkoxysilanes copolymers are currently used as isocyanate-free alternatives for a variety of highly demanding applications, such as marine coatings. Additionally, these systems can, in principle, be formulated as moisture-cure single-component (1K) coatings. However, acrylic-alkoxysilanes exhibit drawbacks, such as slow room-temperature cure and high raw material cost. Despite their high potential, acrylic-alkoxysilanes have had so far only fairly limited market penetration.

This paper presents a new family of polymers, in which vinyl neoester and vinyl alkoxysilane monomers are combined. Variation of process parameters, vinyl monomer composition, and level of alkoxysilane yielded a series of polymers with solids content between 70 and 100%, and exhibiting a broad range of properties. These resins were formulated into 1K coatings and compared with a 2K polyurethane. The relationship between resin processing/composition and resulting coating performance is discussed herein.

Results of this study demonstrate that polymers combining vinyl neoesters and alkoxysilanes can be formulated into 1K clearcoats with shelf lives of at least several months. Unlike when using 2K systems, end users therefore do not need to mix two components shortly before use. Additionally, vinyl silane systems are intrinsically isocyanate-free, and potentially faster drying and with higher solids than 2K polyurethanes. The vinyl-silane systems studied also exhibited excellent weathering resistance.

Keywords: protective silane silicone 1K

S.11-2 Acrylic polyols and polyaspartic esters for isocyanate curing

Mircea Manea
R&D, Alkindo Mitraraya, Tangerang, Indonesia

The awareness of the impact of coatings on the environment has grown almost at the same speed with the market request and coatings production. The driver as finding less polluting solutions has increased along with the environmental concerns in general. This phenomenon has developed on both individual levels as well as through a serious commitment and governmental regulations.

Original contributions are described in relation to synthesis of acrylic polyols. The trials conducted with CAPA monomer as reactive solvent bed in the reactor is leading to acrylic polyols for 2K polyurethane coatings in usual reaction conditions similar to those employed when cooking a trivial alkyd resin. No pressure or solvent is involved in the synthesis which is increasing the yield of the reactor to the 80-100 % filling degree of active volume. The resins exhibit higher reactivity as the hydroxyl functional groups are pushed out of the chain and no sterically hider is involved in the cross-linking reaction. The gloss is also higher as well as pigment wetting properties are improved for equal formulations. This may be a result of the reflow also observed when the use of butane diol mono acrylate is employed in the formulations.

As different from existing aspartic esters based on di-butyl, di-ethyl, di-methyl maleate, and several diamines, poly-functional aspartic esters are described as synthesized from an unsaturated polyester backbone. The design of the polyaspartic has a larger latitude of choice by widening the range of monomers used in the polyester backbone. As well the crosslinking density is a matter of choice depending on the number of maleic anhydride moieties employed in the polyester backbone. This enables a wider formulation possibility when compared to the existing polyaspartic esters which only have two reactive groups.

References:

1. *Contribution of Working Group I to the forth Assessment Report of the Intergovernmental Panel on Climate Change*
2. *Determination of volatile matter content, water content, density, volume solids, and weight solids of surface coatings-Method 24*
3. *EPA website www.epa.gov/ttn/atw/coat/coat.html*
4. *LIST OF EXEMPT SOLVENTS (40CFR51.100 (s), revised as of 7/1/2003)*
5. *J. Brandrup and E. H. Immergut, eds., Polymer Handbook, 3rd ed., Wiley-Interscience, New York, 1989*
6. *H. Lee and K. Neville, Handbook of Epoxy Resins, McGraw-Hill, Inc., New York, 1967, Reprinted 1982.*
7. *W. G. Potter, Epoxide Resins, Springer-Verlag, New York, 1970.*
8. *The Epoxy Resin Formulators Training Manual, The Society of Plastics Industry, Inc., New York, 1984*
9. *C. A. May, ed., Epoxy Resins Chemistry and Technology, 2nd ed., Marcel Dekker, Inc., New York, 1988.*
10. *L. V. McAdams and J. A. Gannon, in J. I. Kroschwitz, ed., Encyclopedia of Polymer Science and Engineering, 2nd ed., Vol. 6, Wiley-Interscience, New York, 1986*
11. *U.S. Pat. 3,947,339*
12. *Wolff, H., Zeidler, G., Rheology of melts and solutions Part 1: viscosity and its measurements, Paint Varn. Prod. Man., July7/August 7(1936)*
13. *Morgan, J.R., Characterization of the cure reactions of bismaleimide composite matrices, Polymer 1997, vol38, (3), pp639-646*
14. *Crivello, R.J., Polyaspartimides: condensation of aromatic diamines and bismaleimide compounds, 1973, vol 11,(6), pp1185-1200*
15. *Browning C. E., Absorption of water in polyimide resins and composites, Polym. Eng. Sci. 18, 16 (1978).*
16. *Apicella A., Nicolais L., Astarita G., Droli E., Apsorption of water in polyimide rsins and composites, Polymer 20(9), 1143 (1979).*
17. *McKague E. L., Reynold J. D., Haskin J. E., Probing the molecular interactions in the diffusion of water in bismaleimide composites, J. Appl. Polym. Sci. 22, 1643 (1978).*
18. *Hawthorn K. I., F. C. Henson, "Epoxy Resin Chemistry II," in R. S. Bauer, ed., ACS Symposium Series No. 221, American Chemical Society, Washington, D.C., 1983,*
19. *Dao, B., et al, Preparation and characterization of some novel bismaleimide monomers and polymers base d on diaminobisimides, High performace polymers, 1996, 8(2), pp243-62*
20. *Liu, Y.L., et al, Novel thermosetting resins based on 4-(N-maleimideophenyl)glycidylether, Polymer, 2004, 45, pp1797-1804*
21. *Li, C., et al., A study of the thermal stability of metal bismaleimides and the thermal cyclodehydration of their precursors at 77K following low temperature irradiation, Cryogenics, 1995, 35(11), pp743-5*
22. *Jin, J.Y., et al, On polyetherimide modified bismaleimide resins, Macromolecular chemistry, 1999, 200, p1956*
23. *Dix L-R., et al, Chain extension and crosslinking of telechelic oligomers, European Polymer Journal 1995, 31(7), p647.*
24. *Liang, G.Z., New bismaleimide resin with improved tack and drape properties for advanced composites, Journal of Applied Polymer Science, 1997, 64, pp273-9.*
25. *Hu, X., Rheological study of crosslinking and gelation in bismaleimide/cyanate interpenetrating polymer network, Journal of Applied Polymer Science, 2001, 80, pp2437-45.*
26. *U.S. Pat. 3,380,964*
27. *Brit. Pat. 1,190,718*
28. *U.S. Pat. 2,818,405*
29. *Stenzenberger H. D., et al. Polyimides, Blackie & Sons, Ltd. 1990. pp79-128*

Keywords: acrylic polyols, polyaspartics, isocyanate curing, synthesis process

S.11-3 Self crosslinking UV curable polyurethane dispersion making waterborne UV technology more efficient

Eva Tejada[1], Michael Ludewig[2], Berta Vega[2]
[1]Application & Development, Covestro SL, Barcelona, Spain,
[2]Covestro Deutschland AG, Leverkusen, Germany

UV curing technologies are the best technology choice when high productivity is needed. As a result, this technology is growing above the market. However, some technical issues are limiting the spread of that technology. Apart from the initial investment needed, UV curable systems have some limitations. On the one hand, it is limited to flat surfaces as it is necessary that light is reaching every single spot of the coating to get the desired result. On the other hand, pigments and high film thickness are also hindering the full curing of the coating. In that article, we will present a solution that will overcome in a great extent that problem, by combining two different cross-linking mechanisms in the polymer. The introduction of alternative cross-linking mechanism to the standard acrylate UV curable functionality is compensating the lack of UV energy and allows good coating performance even if the film is not fully UV-cured. In the product presented here, oxidative curing by means of alkyd modification has been introduced as alternative cross-linking mechanism. This modification has a further advantage: the hydrophobicity of the polymer increases leading to a faster water release.
The combination of UV curing and oxidative curing in the same product improves robustness and efficiency in waterborne UV curable systems, overcoming the main pain points of that technology.

References:

1. *Eva Tejada Rosales, Berta Vega Sánchez, Matthias Wintermantel, "Double crosslinking for greater efficiency", European Coating Journal, November 2017*

Keywords: efficency, self-crosslinking, UV Curing, Water-based coating

S.12-1 Cationic polymerization and network formation of lipase catalyzed vinylether monomers

Sara Brannstrom[1], Maja Finnveden[2], Mats Martinelle[2], Eva Malmström[1], Mats Johansson[1]
[1]Fibre and Polymer Technolog, Stockholm, Sweden
[2]Division of Industrial Biotechnology, KTH: Royal Institute of Technology, Stockholm, Sweden

Photopolymerization is an efficient polymerization method that is very interesting for applications such as photocurable coatings, adhesives, and printing inks. [1] Vinyl ethers (VE) have a strongly electron-donating oxygen atom and consequently VEs has an electron-rich double bond that can be homopolymerized fast using cationic initiators. However, VEs homopolymerize poorly with radical polymerization but mixed with electron/acceptor monomers, like maleates or thiols, they can copolymerize radically in the presence of a photoinitiator upon UV-exposure. [2] Furthermore, the reactive VEs are also known to undergo addition reactions with compounds containing active hydrogens, eg. alcohols and carboxylic acids. [3] Carboxylic acids are attractive monomers due to their vast natural abundance however; making VE-functional esters by traditional esterification is not possible since the acids react with the VE at higher temperatures.

In a previous study a synthetic route to make ester-monomers that contain VE was developed. The monomers were synthesized by lipase-catalyzed acylation of hydroxy vinyl ether with various carboxylic acids. Utilizing mild lipase catalysis it is possible to combine carboxylic acids with hydroxyl vinyl ethers without the occurance of unwanted side reactions due to the acidic carboxylic acid. The monomers are susceptible to wide variations due to numerous readily available carboxylic acids. This permits variation in areas of use and final material properties that can be achieved.

In this work, lipase catalysis is udes under mild conditions to synthesize a series of functional monomers from hydroxyl vinyl ether and various carboxylic acids. The synthesized monomers contain polymerizable vinyl ether functional groups that can be homopolymerized cationically and co-polymerized radically with monomers containing electron poor double bonds.

The use of carboxylic acids as one of the components permits the use of a vast variety of accessible renewable starting compounds with interesting functionalities. Additionally, lipases display chemo, regio and enantio selectivity, which can be used to add more value if more complex monomers are desired. In previous work, the reaction between the hydroxy vinyl ether 1,4-butanediol vinylether (BDVE) and 10-undecenoic acid (UA) was studied in detail in terms of solvent and temperature compatability. Additionally, the kinetics of radical and cationic polymerization of the bifunctional monomers was studied with real time FTIR. In this study the aim is to study the polymerization in more detail and make crosslinked films with a varying crosslinking density. The vinyl ethers are primarily polymerized with UV-initiated cationic polymerization.

References:

[1] S. Rajaraman, W. Mowers, and J. Crivello, Macromolecules, 1999, 32, 36-47
[2] C. Decker, C. Bianchi, D. Decker and F. Morel, Progress in Organic Coatings, 2001, 42, 253-266
[3] T. Mukaiyama and H. Nohira, Bulletin of the Chemical Society of Japan, 1960, 33, 1716-1720

Keywords: cationic polymerization, enzymatic catalysis, photopolymerization, vinylethers

S.12-2 One-step synthesis of novel bio-based flexible polyurethane coatings using unmodified kraft lignin

Stephanie Elisabeth Klein[1], Margit Schulze[1], Sandra Brück[1], Steffen Witzleben[1], Matthias Rehahn[2]
[1]Dept. of Natural Sciences, Bonn-Rhein-Sieg University of Applied Sciences, Rheinbach, Germany
[2]Dept. Macromolecular Science, Technical University Darmstadt, Darmstadt, Germany,

Motivation

Lignin is a valuable by-product of the paper and pulping industry, available in industrial amounts in form of so called black liquor mainly produced via Kraft pulping

- [1]. Hitherto the utilization is limited to energetic use. Lignin consists of three monolignol units containing aliphatic and aromatic hydroxyl groups that determine the chemical reactivity and enable polyurethane formation
- [2]. Due to the importance of polyurethanes in construction industry, lignin-based polymers gained increasing interest [3, 4].

Aims

The present work is focused on extraction of lignins from hard wood and subsequent utilization of the isolated and purified lignin as a substitute for fossil-based expensive diol (polyol) components in polyurethane synthesis.

Material and methods

Kraft lignin is extracted from industrial black liquor via acidic precipitation using H^2SO^4 [4]. FTIR spectra of Kraft lignin show characteristic OH bands and CH signals in the fingerprint region. Three distinct UV-Vis absorptions are observable at 220 nm, 245 nm, and 300 nm. For the One-Step synthesis of duroplastic polyurethanes the Kraft lignin was added to a diisocyanate, tetrahydrofuran, plasticizer and triethylamine mixture. Titration according to ISO 14900:2001(E) and ^{31}P-NMR studies were performed to determine the hydroxyl number of Kraft lignin. The NCO:OH ratio and the pH level of Kraft lignin were varied to modify the mechanical properties of the resulting polymers.

Results

Lignin-based highly flexible and temperature stable polyurethane coatings were obtained. Characterization was performed including: a. ATR spectroscopy to confirm the formation of urethane bonds (N-H stretching at 3382 cm^{-1}, C=O stretching at 1620 cm^{-1}); b. swelling behavior in different solvents confirming that lignin is completely incorporated in the polyurethane network; c. scanning electron microscopy (SEM) and thermal analysis (TGA) showing highly flexible thin (d = 159 µm) and temperature resistant (150 - 170°C) bio-based PU coatings.

Acknowledgement

Financial support is given Federal Ministry of Eduction and Research (BMBF) program IngenieurNachwuchs, project LignoBau (03FH013IX4).

References:

[1] Hemmilä V et al. RSC Adv 2017, 7, 38604-38630
[2] Griffini G et al. ACS Sustainable Chem Eng 2015, 3, 1145–1154
[3] Hansen B et al. J Polym Environ 2016, 24, 85-97
[4] Witzleben S et al. J Chem Chem Eng 2016, 9, 494-499

Keywords: lignin, polyurethane coating

S.12-3 Sugar-based polyesters: from enzymatic synthesis to coating application

Liliana Gustini[1, 2, 3,] Cristina Lavilla[3, 4,] William W. Janssen[3], Antxon Martinez de Ilarduya[5], Sebastián Muñoz-Guerra[5], Cor E. Koning[3, 4]
[1]Sappi Netherlands Service, Maastricht, the Netherlands
[2]Dutch Polymer Institute, 3Laboratory of Physical Chemistry, Eindhoven University of Technology, Eindhoven, the Netherlands
[4]DSM Coating Resins, Zwolle, the Netherlands
[5]Departament d'Enginyeria Química, ETSEIB, Universitat Politècnica de Catalunya, Barcelona, Spain

Renewable polyesters derived from a sugar alcohol, i.e. sorbitol, were synthesized by solvent-free polycondensation. The target was to obtain linear polyesters with pendant hydroxyl groups along the polymer backbone. The performance of the biocatalyst SPRIN liposorb CALB and the organo-base triazobicyclodecene (TBD) were compared with two metal-based catalysts, viz. scandium trifluoromethanesulfonate (Sc(OTf)3) and dibutyl tin oxide (DBTO). For the four catalytic systems, the efficiency of the incorporation of sorbitol and the selectivity towards the primary versus secondary hydroxyl groups were studied, mainly via 13C and 31P NMR spectroscopy while the occurrence of side reactions was evaluated via MALDI-ToF-MS. The incorporation of sorbitol was especially successful with SPRIN liposorb CALB, achieving high selectivity without side reactions and leading to close-to-linear polyesters 1. Subsequently, polyesters based on renewable sorbitol and isohexides building blocks, viz. isomannide as diol and isoidide dimethyl ester as acyl-donor, were obtained via solvent-free lipase-catalyzed polycondensation. Pendant hydroxyl groups coming from the sorbitol units were present along the polyester backbone, whereas the two isohexides introduced rigidity into the polyester chains. Remarkably, while CALB incorporates isomannide through secondary hydroxyl groups, the selectivity between primary and secondary hydroxyl groups in sorbitol was maintained, offering a unique strategy to tailor the molecular structure of the polymer. The presence of bicyclic units enhanced Tg with respect to the parent sorbitol-containing polyester lacking isohexides. Solvent-borne coatings were prepared by cross-linking the sugar-based polyester polyols with di- and triisocyanates. The increased Tg of the isohexide containing polyesters led to an enhancement in hardness of the resulting coatings. This study highlights the benefits of enzymes as selective catalysts in polycondensation reactions, and the application of sugar-based polyesters in coatings 2.

References:

[1] L. Gustini, C. Lavilla, W.W.T.J. Janssen, A. Martinez de Ilarduya, S. Muñoz-Guerra, C.E. Koning ChemSusChem. 2016, 9, 2250.
[2] L. Gustini, C. Lavilla, W.W.T.J. Janssen, A. Martinez de Ilarduya, S. Muñoz-Guerra, C.E. Koning Biomacromolecules 2016, 17, 3404.

S.13-1 Pushing the boundaries of waterborne two component non-isocyanate coatings.

Dirk Mestach, Silfredo Bohorquez
Research & Development, Allnex Netherlands BV, Bergen op Zoom, the Netherlands

Waterborne two component polyurethane coatings have been commercially available for the last 25 years. Over the years, a lot of progress has been made in order to improve the final properties of these coatings, but as of the fact that the crosslinking reaction depends on the reaction of aqueous polyols with emulsified or dispersed polyisocyanates, the layer dependant formation of surface defects that arise from reaction of isocyanate with water is still a concern. Furthermore polyisocyanate crosslinkers are under scrutiny from legislators, so their longer-term use might become an issue.

Around the same time as the introduction of waterborne two component polyurethanes, Allnex developed a non-isocyanate ("NISO") binder system onto the market. The system was based on an epoxy-functional polymer dispersion made by means of emulsion polymerization that was combined with a secondary emulsion of an acrylic polymer carrying bot carboxylic acid and tertiary amine functional groups. This first generation binder system offered reasonable performance and was/is used extensively in both metal and wood-coatings. Some significant drawbacks were the high production cost for the secondary emulsion and the limited stability and pot-life after mixing both components.

Recently some significant breakthroughs have been made that have led to a next generation two pack non-isocyanate system. In this system both components are now made by emulsion polymerization and some of the drawback of the first generation system were solved.

In this paper the cross-linking kinetics are discussed together with the mode of film-formation and the way how they influence final performance in an industrial woodcoatings application. Techniques used include FT-IR and environmental scanning electron microscopy.

Keywords: cross-linking, film-formation, non-isocyanate, two component

S.13-3 Novel oil-modified acrylic multi-domain dispersions

Andreas Steiner[1], Leila Khatai[2], Tim Kittler[3], Luc Sterckx[4], Kimberly Gariepy[5], Florian Lunzer[1]
[1]Research, Allnex Austria GmbH, Graz, Austria,
[2]Marketing, Allnex, Drogenbos, Belgium,
[3]Marketing, Allnex, Alpharetta, USA,
[4]Technical Service, Allnex, Drogenbos, Belgium,
[5]Technical Service, Allnex, Alpharetta, USA

Waterborne acrylic dispersions and alkyd emulsions have been extensively used for architectural coatings for decades, each technology having its advantages and drawbacks. While alkyd emulsions generally exhibit very good application characteristics such as high gloss and good penetration, they lack in performance for weathering resistance. Acrylic dispersions, conversely, are disadvantaged in terms of flow, leveling and appearance, but perform well for outdoor durability. Many attempts have been undertaken to find the balance of performance by hybrid technology approaches. The presented work highlights a novel synthetic approach to oil modified acrylic multi domain dispersions combining acrylic-like outdoor durability with alkyd-like appearance and application behavior.
This innovative synthetic concept, along with supportive performance testing results, will be outlined and discussed.

Keywords: acrylic dispersion, multi-domain, oil-modified, wood coating

S.14-1 Tailoring of versatile surface morphologies on hot dip galvanized steel in wet supercritical CO^2: formation, corrosion resistance and properties as a substrate for coatings

Ville Saarimaa[1], Aaretti Kaleva[2], Juha-Pekka Nikkanen[2], Jere Manni[1], Carl Lange[1], Teemu Paunikallio[1], Saara Heinonen[2], Erkki Levänen[2], Pasi Väisänen[3], Golrokh Heydari[4], Antti Markkula[3]
[1]Top Analytica Oy Ab, Turku, Finland
[2]Laboratory of Materials Science, Tampere University of Technology, Tampere, Finland
[3]SSAB Europe Oy, Hämeenlinna, Finland,
[4]SSAB Europe Oy, Stockholm, Sweden

In outdoor exposure, zinc coatings undergo complex corrosion processes depending on the prevailing conditions. Typically, the first corrosion products (white rust: $Zn(OH)^2$ and ZnO) appear on unprotected zinc coatings within some days or weeks, while formation of more stable forms of patina (zinc hydroxy carbonates) can take several months or years. The stable, impurity-free zinc patina increases the electrochemical stability of zinc and promotes adhesion to organic coatings. However, the reaction products of zinc have so far not been utilized as zinc pretreatments for coating processes, since formation of such layers in outdoor exposure is uncontrolled and time-consuming. Furthermore, the naturally formed patina is easily contaminated by impurities, lacks specific morphology and is subject to Zn run-off during the exposure. The main purpose of this study was to develop a novel, environmentally friendly surface treatment for hot dip galvanized steel that features good electrochemical stability and provides a good substrate for coatings. Carbon dioxide is abundantly available and cost efficient. Wet supercritical carbon dioxide ($scCO^2$) was used as a reaction medium for rapid patina formation. The presence of H^2O causes formation of carbonic acid, which accelerates zinc dissolution and promotes the formation of Zn corrosion products. Uniformity of the patina layer was further improved using a Cu^2+ precursor that catalysed zinc oxidation. Two distinct patina forms were obtained by adjusting the treatment temperature: $ZnCO^3$, and zinc hydroxycarbonate nanowires. The patina forms could be formed separately or as a hybrid layer. Measurement data obtained with various analytical tools (FTIR, RAMAN, XPS, SEM) is presented to show the composition and morphology of the structures. The wet $scCO^2$-induced patina improved corrosion protection and provided a good substrate for coating. Scanning vibrating electrode technique (SVET) showed anodic currents (active zinc) on zinc surface, and neutral to cathodic currents (passive zinc) on patina-covered surface. Passivation layers are inert layers that inhibit interaction of metallic zinc with atmosphere, but rarely function as primers for organic coatings. Pretreatments, in turn, provide very little corrosion resistance as such, but are essential for good adhesion of organic coatings. $ScCO^2$-induced zinc patina has the potential to combine the assets of both passivation layers and pretreatments.

Keywords: corrosion resistance, galvanized steel, patina, supercritical carbon dioxide

S.14-2 Application of hydrophobic ultra-low viscosity HDI polyisocyanate for water-borne 2K PU industrial coatings and polyaspartic coatings

Satoshi Takeno, Junya Tsujita
Performance Coating Materials R&D Department, ASAHI KASEI Corporation, Kawasaki-city, Japan

This presentation will show how low viscosity HDI polyisocyanates will perform in water-borne 2KPU and polyaspartic application.

The use of hydrophobic polyisocyanates in water-borne 2KPU applications usually lead into disadvantages because of poor compatibility. That is also a reason why high amounts of co-solvents are requested. Thereof our original ultra-low viscosity HDI polyisocyanate has been developed to bridge all these gaps.

It will be shown that our ultra-low viscosity HDI polyisocyanate gives coating formulating companies the opportunity to create water-borne 2KPU systems based on a hydrophobic polyisocyanate with reduced co-solvent and improved optical properties and keeping good mechanical as well as chemical resistance performance of the related film.

In addition, when the ultra-low viscosity HDI polyisocyanate is applied to the polyaspartic coating system, coating performance such as drying property, hardness, and weather resistance can be maintained while lowering the viscosity of the formulation (while attaining VOC reduction) than the standard trimer. This is a result of the trimer structure being the main component in this grade. Furthermore, when it is desired to make a solvent-less formulation, our ultra-low viscosity HDI polyisocyanate is considered to be optimum.

S.14-3 Easy-to-clean additives for today and tomorrow

Majdi Al-Masri, Volker Thyssen-Wallner
BYK Chemie GmbH, Wesel, Germany

Expressions such as anti-dirt, anti-stain, anti-soil, anti-graffiti, easy-to-care, and easy-to-clean are meanings for virtually the same phenomenon: surface repellency of solid powder particles and liquids.
An easy-to-clean effect can be achieved by creating high hydrophobic or high hydrophilic surfaces and by using structure-giving particles or by preparation of surfaces. For an additive producer it is very challenging to develop an additive for different formulations irrespective of the chemistry of the coating system.
This paper will illustrate the stages of the easy-to-clean hydrophobic additives developed in the last decade and provide an explanation for their behavior in relation to the chemistry involved.
The paper will also provide a summary of the existing easy-to-clean additives and describes the benefits and limits of each system. It will illustrate the suitable coating systems for each additive and demonstrate the other effects of the additives on corresponding systems. Results which were achieved by conducting application tests using colored markers, powders, and slurries will be presented. In addition to standard coating tests like contact angle, surface tension and surface energy measurements, results of simulated long-term tests will also be featured. New light will be shed on some current research results and an overview of future developments and challenges will be given.

Keywords: silicone, additive, surface, easy-to-clean

S.15-1 A quantum leap in ski waxing technology

Jan Inauen[1], Konstantin Siegmann[1], Diego Villamaina[2], Robert Sterchi[1], Martin Winkler[1]
[1]IMPE, ZHAW, Winterthur, Switzerland
[2]University of Zurich, Zurich, Switzerland

Faster and faster - this also applies to skiing. In order to meet this requirement, good ski wax is needed. Ski wax minimizes the sliding resistance of the skis on snow. The friction of the skis on the snow forms a thin film of water, which lubricates and is responsible for the gliding of the skis. The more water-repellent the surface, the faster the ski. Today's highly fluorinated carbon compounds are quite effective in increasing hydrophobicity. All ski waxes, however, have in common that they are quickly rubbed off and hence friction increases. Conventional ski wax does not even adhere sufficiently long for the duration of one race. It would be desirable to use a wax that forms a firm bond with the surface of the skis.
The challenge is that the ski's running surface is made of ultra-high molecular weight polyethylene (UHMWPE), a very inert plastic. The so-called grafting of UHMWPE requires drastic reactions such as those known from nitrenes. A nitrene is a nitrogen species with only six valence electrons, a sextet, which is highly reactive and can therefore insert into C-H bonds. If it were possible to couple the wax to a nitrene, the wax could thus be attached to the ski's surface. Photoreactive ski wax was designed according to this idea.
As graft chains perfluoroalkanes were chosen, and UV-light induced grafting was selected as grafting method. Hydrophobic photoreactive surface modifiers (ski wax) based on azide/nitrene chemistry were designed, synthesized in high yield and characterized. Four new molecules were described. Using spectroscopy it was shown that, indeed, a monomolecular, covalently bound grafted layer is formed from the photodecomposition of a perfluorophenyl azide on polyethylene surfaces.
A model compound mimicking the grafted nitrene species was synthesized. Comparison of the UV-spectra of this compound to the one from the photografted and washed layer indicated the formation of a monomolecular coating. X-ray photoelectron spectroscopy (XPS) of the nitrogen species from the photografted layer showed a peak whose binding energy was comparable to the one of the model compound.
An abrasion test was designed. A rotating polyethylene disc, coated with ski wax, was dipped into a water bath with small pieces of ice. The ice collides with the disc and rubs it off, similar to a skiing trip. The abrasion of the ski wax could be determined by means of XPS as a function of time. It was shown that photoreactive ski wax is more than twice as durable as conventional ski wax. Extensive field tests with professional ski racers followed. In cross-country skiing, for example, the time saving was 0.1 - 0.3 seconds in a short descent of 20 seconds, which correspondents to an improvement of 0.5 - 1.5%. This by far exceeds the performance improvement of 0.5% demanded at the beginning of the project, so that one can truly speak of an actual quantum leap in ski waxing technology.

Keywords: grafting, hydrophobic, polyethylene, ski wax

S.15-2 Sol-gel encapsulation of microorganism: coating characterization and enzyme activity determination

Tangi Senechal[1], Driss lahem[1], Cristiana Cordeiro De Castro [2], Anne-Lise Hantson[2]
[1]SDM, Materia Nova, Mons, Belgium
[2]UMons, Mons, Belgium

Indoor environmental contamination is a major problem that needs efficient, durable and environmentally friendly solutions. The development of active coatings able of entrapping and degrading selectively pollutants is of great interest. In the frame of the developpment of new coatings for the purification of indoor air, the encapsulation of microorganism for formaldehyde degradation into sol-gel and latex coating have been studied. The conservation of the enzyme activity is the key parameter for the long-term efficiency of this kind of coating. Sol-gel matrices present good mechanical resistance, durability and easy application but need some adjustement in order to keep enzyme activity. Formulations and applications must be adapted avoiding mechanical stresses, high post curing temperature, contact with solvent and extreme pH. The determination of enzyme activity on applied coating is also discussed in this presentation. Pleutorus ostreatus is used as a model microorganism allowing to follow laccase activity when encapsulated into the sol-gel coating. This paper will present the sol-gel preparation and microorganism incorporation in order to keep enzymatic activity, like PVOH addition for room temperature reticulation and flexibilty or rotavapor step to remove solvent. The characterization of laccase activity by ABTS reactive is conducted on the coated substrate which represent more difficulties than directly in solution, but is more representative of the final application. The influence of different sol-gel precursor ratio, microorganism ratio and type of application on the laccase activity will be studied. The study of water availability and diffusion through the coating will also be a key parameter to preserve a long term enzymatic activity. The activity of microorganism encapsulated in sol-gel will be compared to a latex encapsulation.

References

1. A. Coiffier, T. Coradin, C. Roux, O. M. M. Bouvet and J. Livage, J. Mater. Chem., 2001, 11, 2039–2044.
2. N. Nassif, C. Roux, T. Coradin, M.-N. Rager, O. M. M. Bouvet and J. Livage, J. Mater. Chem., 2003, 13, 203–208.
3. R. B. Bhatia, C. J. Brinker, A. K. Gupta and A. K. Singh, Chem. Mater., 2000, 12, 2434–2441.
4. S. Fennouh, S. Guyon, J. Livage and C. Roux, J. Sol-Gel Sci. Technol., 2000, 19, 647–649.
5. J. Rajan Premkumar, R. Rosen, S. Belkin and O. Lev, Anal. Chim. Acta, 2002, 462, 11–23.
6. J. R. Premkumar, O. Lev, R. Rosen and S. Belkin, Adv. Mater., 2001, 13, 1773–1775.
7. M. L. Ferrer, L. Yuste, F. Rojo and F. del Monte, Chem. Mater., 2003, 15, 3614–3618.
8. N. Nassif, O. Bouvet, M. Noelle Rager, C. Roux, T. Coradin and J. Livage, Nat. Mater., 2002, 1, 42–44.

Keywords: enzyme activity, microorganism, sol-gel coating

S.15-3 Corrosion control by self-assembled nanolayer coatings

Talah Abohalkuma, Abdul Shaban, Judit Telegdi
Research Centre for Natural Sciences, Budapest, Hungary

The control of undesired corrosive deterioration can be achieved by different techniques. All take aim of insulation of a metal surface from the aggressive environment. It could happen by addition of chemicals (inhibitors) to the aqueous solution, by application of coatings (of different origin and used in different thickness) as well as by molecular layers. The type of methods depends on the function. Our research has aimed at the application of self-assembled molecular layers (SAMs) that can control the corrosive processes as well as the microbial adhesion of corrosion relevant microorganisms. The advantage of this technique is the use of chemicals at very low quantity (to decrease the undesired environmental impact), and the simplicity of the method. In our experiments, two amphiphilic molecules with phosphonic head-groups formed the self-assembled nanolayers. There are significant differences between the two hydrophobic chains. One is more hydrophobic than the other is, but the other one contains a double bond that is capable for polymerization. The most proper SAM preparation technique as well as the characterization of the nanolayers by surface analytical and visualization methods will be demonstrated. The time dependent molecular layer formation was correlated with the change in the wettability that represents the compactness of the layer. It is in correlation with the anticorrosion activity, which was followed by electrochemical techniques and by atomic force microscopy. The change in the surface roughness characterizes the corrosive processes. The electrochemical impedance spectroscopy gives information on the mechanisms of the corrosion. By electrochemical polarization, the type of the inhibition is demonstrated. In order to increase the anticorrosion activity of the molecular layers the well-ordered layer structure was modified by follow-up treatments.

References:

1. *J. Telegdi, G. Luciano, S. Mahantry, T. Abohalkuma; Inhibition of aluminum alloy corrosion in electrolytes by self assembled fluorophosphonic acid molecular layer; Materials and Corrosion, 67 (2016)1027-1033.*
2. *T. Abohalkuma, A. Shaban, J. Telegdi; Corrosion Processes Controlled by Phosphonic Acid Nano-layers; Periodica Polytechnica Chemical Engineering 60 (2016)165-168.*
3. *J. Telegdi, L. Trif, L. Románszki; Book Chapter 9: Smart anti-biofouling composite coatings for naval applications in: "Smart composite coatings and membranes: Transport, structural, environmental and energy applications". Woodhead Publishing, (2016) pp. 123-155. ED: Montemor MF (ISBN:978-1-78242-283-9).*
4. *T. Abohalkuma, J. Telegdi; Corrosion protection of carbon steel by special phosphonic acid nano-layers; Materials and Corrosion 66 (2015) 1382-1390.*
5. *T. Abohalkuma, F. Shawish, J. Telegdi; Phosphonic acid derivatives in self-assembled layers against metal corrosion; International Journal of Corrosion and Scale Inhibition 3 (2014) 151-159*

Keywords: anticorrosion nanocoatings, self-assembled molecular layers

S.16-1 Behind the design rules to develop self-healing coatings

Santiago J. Garcia
Novel Aerospace Materials, Delft University of Technology, Delft, the Netherlands

Coatings are generally applied on bulk materials and structures with the main intention of implementing properties and functions that the underlying substrate lacks. Such functions can be as diverse as aesthetics, corrosion protection, hydrophobicity or hydrophilicity, drag reducing or anti-icing. Unfortunately, sudden or time-dependent damage in coatings typically leads to the loss of their main function and ultimately to the need of structure replacement. As a way to reduce the impact of damage, diverse intrinsic and extrinsic self-healing strategies have been proposed in the last years. The vast amount of proposed chemistries, carrier systems and inhibitors in the case of anticorrosive coatings is a clear sign of the broad interest this field has attracted. Our contribution to the field in the last years has focused mainly on establishing key governing rules for a better design of self-healing polymers and coatings as well as on the development of healing strategies up to the proof-of-concept level.

In this presentation I will show some of our ongoing efforts to establish design rules to develop self-healing polymeric and anticorrosive coatings and how they led to new and improved concepts. Special attention will be put on three aspects: (i) the relation between polymer architecture, dynamics and healing in brush polymers and thermosets; (ii) the design of novel techniques to better understand and visualize the healing phenomena such as laser speckle interferometry; and (iii) the potential of bio-inspiration to develop more efficient anticorrosive coatings.

Keywords: bioinspired, corrosion protection, new techniques, self-healing coatings

S.16-2 Self-healing waterborne polymer coatings

Sil Nevejans[1, 2], Nicholas Ballard[1], Bernd Reck[2], José M. Asua[1]
[1]POLYMAT, University of the Basque Country UPV/EHU, Donostia-San Sebastián, Spain,
[2]Dispersions and colloidal materials research, BASF SE, Ludwigshafen, Germany

Recently, self-healing polymers based on disulfide compounds have gained attention due to the versatile chemistry of disulfide bonds and easy implementation into materials. The increasing interest in self-healing poly(disulfide)s have not only led to the clarification of the underlying mechanisms of disulfide exchange reaction,[1,2] but also to the development of self-healing materials based on disulfide bonds which can even heal at room temperature without the use of any external factors or catalysts.[3]

In this work, the development of intrinsically healable polymer films useful for waterborne coating application based on the incorporation of these attractively dynamic disulfide bonds will be presented. Polyurethanes were selected to serve as backbone for these materials, as they generally provide very good physical properties for coatings. Although the mechanical strength of polyurethanes is appealing for their use in coatings, it introduces at the same time one of the most important challenges for designing self-healing coatings. The strength of the material, which is indispensable for coating applications, implies a restricted motion of the polymer network, while on the other hand network mobility is critical for the materials ability to repair any damages and self-heal.

Here, we will focus on the synthesis of waterborne poly (urea-urethane) s (PUUs) dispersions which could be used to produce a healable coating that might be able to fulfill the aforementioned opposing requirements. In other words, the aim is to create a material that is hard enough to protect a surface, but additionally provides a certain mobility to recover from damage. First, the synthesis procedures of these waterborne PUU dispersions will be discussed, after which the influence of different reactants on the polymer characteristics will be demonstrated. Finally, various qualitative and quantitative experimental techniques will be used to evaluate the healing performance of the intrinsically healable PUUs and to test its usefulness in coating applications.

References:

1. *Matxain, J. M., Asua, J. M. & Ruipérez, F. Design of new disulfide-based organic compounds for the improvement of self-healing materials. Phys. Chem. Chem. Phys. 18, 1758–1770 (2016).*
2. *Nevejans, S., Ballard, N., Miranda, J. I., Reck, B. & Asua, J. M. The underlying mechanisms for self-healing of poly(disulfide)s. Phys. Chem. Chem. Phys. 18, 27577–27583 (2016).*
3. *Rekondo, A. et al. Catalyst-free room-temperature self-healing elastomers based on aromatic disulfide metathesis. Mater. Horizons 1, 237–240 (2014).*

Keywords: dynamic disulfide bonds, intrinsic self-healing, waterborne coating

S.16-3 Cross-linked composite polyrotaxanes for self-healing coatings

Nina Müller[1], Jessica Brunke[1], Budiman Ali[1], Gerhard Wenz[2], Peter W. de Oliveira[1], Carsten Becker-Willinger[1]
[1]INM Leibniz-Institute for New Materials, Germany
[2]Saarland University, Saarbrücken, Germany

Self-healing surfaces are desired, e.g. for car paints and smartphone covers. Different approaches of intrinsically self-healing materials are known. [1] Thermally reversible crosslinked polymers are well-established but usually needs high temperature for healing. A physical approach toward self-healing is based on a stress relaxation mechanism. Slide-ring gels (SRGs), which could be composed of cross-linked polyrotaxanes, are such materials. [2,3] Polyrotaxanes (PRs) are supramolecular assemblies which comprise a ring-shaped molecule and a polymer. The ring-shaped molecules are threaded onto the polymer like beads on a necklace. Usually, these PRs are prepared in a multi-step synthesis via threading ring-shaped molecules on a linear polymer and introducing bulky groups to prevent the rings from sliding of the backbone. Recently a new one-pot emulsion polymerization was developed where monomers are complexed by cyclodextrins (CDs) and afterwards co-polymerized with a bulky co-monomer. [4,5] With this new method, larger amounts of PRs are available for the development of coatings and paints.
The threaded CD molecules provide hydroxyl groups which can be used for cross-linking. By cross-linking the ring-shaped molecules a SRG is achieved. Unlike in traditional polymer networks the chains are not directly cross-linked but connected via the cross-linked ring-shaped molecules. SRGs are highly flexible as the cross-linked groups can freely move over the chain when a force is applied. Therefore, the polymer can draw aside when a scratch is applied and permanent damage can be avoided. Small scratches based on displaced material can self-heal when the coating is heated above glass transition temperature. [2,3]
Lacquers of PRs and bis- or polyisocyanates as cross-linking agent can be applied via spin or spray coating. In some cases additives for better film quality or UV stability or modified nanoparticles for improved mechanical properties are added. The coatings are thermally cured to achieve cross-linking. The resulting smooth and transparent coatings achieve Martens hardnesses over 200 MPa which makes them comparable to conventional clear coats. The flexibility of the material can be tuned over the sliding length of the rings, i.e. the amount of co-polymerized stopper groups. The pencil hardness (Wolff-Wilborn) of the coating is 1H. Due to the mobility of the ring-shaped molecules, artificial damages with a depth of 3 µm can be healed completely within 1 minute upon heating up to 100 °C. Dynamic-mechanical-thermal analysis revealed a glass-transition around the self-healing temperature. [6]
In summary, cross-linked PRs based on CDs as ring-shaped molecules provide good mechanical properties and self-healing behavior and therefore have a great potential for applications as top coat for technical devices or cars.

References:

1. *Wu D.Y., Meure S., Solomon D., Prog. Polym. Sci. 2008, 33, 479*
2. *Kato K., Matsui D., Mayumi K., Ito K. Beilstein J. Org. Chem. 2015, 11, 2194*
3. *Araki J., Kataoka T., Ito K., Soft Matter 2008, 4, 245*
4. *Kali G., Eisenbarth H., Wenz G. Macromol. Rapid Commun. 2016, 37, 67*
5. *WO 2016/202906 A1*
6. *EP16205619.6, not published*

Keywords: composite, self-healing coatings, slide-ring gel

S.17-1 Inkjet printing - a technology for coating industry

Nicolae Tomozeiu
Océ Technologies B.V., Venlo, the Netherlands

Printing is a special coating technology that in the last decennia had a tremendous development. It means in this context, deposition of a functional layer on a substrate using the digital printing. Nowadays the world of printing is rapidly evolving, especially due to the inkjet technologies, which offer the flexibility of digital printing at a breakthrough cost price, productivity, print quality and versatility.

The R&D department of Océ Technologies, a Canon company, is a major player in the development of inkjet technologies for many different applications. An overview of the Océ inkjet technologies is given in this work, including strengths and limitations. The main focus will be on printing coatings with water based inks that contains polymeric particles. When such an ink droplet meets a substrate, physical processes as spreading, evaporation, absorption and hardening of the polymeric component take place. Understanding them is vital to have coatings of high quality; and this is the aim of this work, to develop and share knowledge. The influence of the liquid physical properties as well as of the substrate characteristics on the coatings hardening will be considered. Theories and results of experimental studies based on optical spectroscopy, microscopy, Scanning Electron Microscopy (SEM), Nuclear Magnetic Resonance (NMR) and Automatic Scanning Absorptometer (ASA) are presented. For each method of investigation we will present the main models, including their strengths and limitations.

The science of printing is at home within Océ Technologies, a Canon company. The future of inkjet printing - a technology for coating industry, is excellent.

Keywords: inkjet, optical spectroscopy, printing as a coating technology, waterborne coating

S.17-2 Everything under control? Substrate wetting additives for waterborne coating and ink formulations

Susanne Struck[1], Ingrid Meier[2], Fabian Eichenberger[1]
[1]Applied Technology, Evonik, Coating Additives, Essen, Germany,
[2]Innovation Management, Evonik, Coating Additives, Allentown, USA

This paper takes scientists and formulators on a tour through the world of surfactants, interfaces and phenomena at the coating/substrate interface. By creating a general understanding of the properties of the different wetting agent classes and relating them to practical applications, the authors provide guidance for easier selection of the most suitable products.
The authors give an introduction into substrate wetting and related phenomena, such as flow, substrate wetting, anti-cratering, etc. They describe the different surfactant classes commonly used.
A comparative study on various different surfactants has been performed on dynamic and static (also known as equilibrium) surface tension, foam behavior, substrate wetting and anti-crater properties.
Through these findings, as well as practical application results, the authors discuss structure/property relationships and highlight the importance of static versus dynamic surface tension in different applications.
In a nutshell the authors come to the following conclusions:

- Substrate wetting additives are not only efficient problem solvers, but significantly enlarge the process window for a coating or ink.
- No wetting agent exists that is the perfect solution for every problem. Instead, there is a wide range of effective organic and siloxane-based wetting additives that cater to different needs.
- When choosing a substrate wetting additive the formulator needs to consider surface tension reduction as well as performance in the actual application, where the ability of a wetting agent to interact with other system components and to affect other properties becomes critical.
- Because of their structures, organic and siloxane-based gemini surfactants display unique performance in coatings and inks, generally requiring lower dosage levels and having lower foaming tendencies than traditional surfactants.

Keywords: Additives, organic and siloxane-based surfactants, Substrate wetting, surface tension

S.17-3 1K hydrophobic binders for anti-corrosion primers and DTM topcoats

Thomas Bernhofer, Jan Pilger
Technical Service Coatings, Synthomer Deutschland GmbH, Marl, Germany

Metal corrosion creates each year an economic damage of several billion USD. To maintain the aesthetical aspect and the value of expensive goods Synthomer developped a dedicated product range for long lasting metal protection. Furthermore these waterborne binders are environmentally friendly, due to low VOC content and being free of hazardous components like APEO and formaldehyde.

The binders are especially designed for application at ambient or slightly forced temperature. Typical industry end-use for these binders is steel frame construction (e.g. power pylons) or ACE (agriculture, construction and earth moving equipment). Achievable anti-corrosion categories according to EN 12944-2 are C 3-C 4 = medium protection level (5 – 15 Y). Hydrophobicity as key parameter for good corrosion protection was optimized by applying modern test-methods like contact angle measurements or electrochemical impedance.

References:

1. *Christ, U.; Nothelfer-Richter, R. Farbe und Lack. 2016 (05), 22-23*
2. *Gagro, D. Farbe und Lack. 2013 (12), 12-15*
3. *Jones, D. A. Principles and prevention of corrosion; Prentice Hall: Upper Saddle River, 1996.*
4. *Contact angle. https://www.kruss.de/de/service/schulung-theorie/glossar/ /kontaktwinkel*
5. *University of applied sciences Niederrhein, Report experimental set-up electro-chemical impedance*

Keywords: anticorrosion primer, DTM-coating, electrochemical impedance, hydrophobicity

S.18-1 A novel approach to develop surface-active antimicrobial coatings

Barend van de Velde, Hossein Mahmoud
Croda, Gouda, the Netherlands

The possibility that bacteria and or other micro-organisms such as fungi, molds and algae accumulate on surfaces is a threat to the durability of materials as well as to human and animal health. This problem is evident on many materials and in many situations, such as medical devices, ship hulls and any moist environments in house- holds, e.g., showers, bathrooms, and kitchens. Over the years many ways of tackling this problem have been developed and have become well known in the art. Most available antimicrobial coatings are active by a slow release of toxic biocides like silver-ions to its environment. The slow release and accumulation of these biocides is of growing concern. The Regulation of Antimicrobials in paint and coatings are getting stricter and are driving the industry towards the development of more environmentally safe antimicrobials. The environmental impact of the leachable antimicrobials from façade coatings is one of the examples which have attracted increased attention in recent years, while the microbial resistance to these antimicrobials is another major challenge.

In order to achieve long-active antimicrobial solutions Croda develops non-toxic, surface-active antimicrobial resins for paint applications. By using a special polymerization process, an antimicrobial resin is fabricated. This resin develops a cationic/ hydrophobic surface during the curing (hardening) process. A polymer with different charges and polarity ranges on its surface can induce electrostatic attraction to the different parts of bacterial cell membranes, making it a hostile environment and thus reducing their proliferation. Like using a mouse trap instead of a poison, the novel technology works like a kind of microbe trap on a nano scale. Apart from being completely safe for man and environment, this physical action has another big advantage: microbes will not become resistant to this kind of control. The antimicrobial technology, its working mechanism and test results will be discussed.

Keywords: Antibacterial coating, binder, biocide, microorganism

S.18-2 Preparation of microencapsulated mat-PDMS and use them in self-healing epoxy coatings

Morteza Ebrahimi[1], Amir Khalaj Asadi[2], Mohsen Mohseni[2]
[1]Polyemer Eng. and Color Tech, Tehran, Iran
[2]Amirkabir University of Technology, Tehran, Iran

One of very effective approaches in the field of self-healing systems is to use healing agents which can be cured in natural conditions (i.e. in the presence of atmospheric moisture, oxygen or sunlight). Photo-responsive self-healing systems is also particularly attractive because they are clean, rapid, remote, cheap, readily available and practical in strategy.

In this work, we have developed an extrinsic photoinduced self-healing system for metal coatings application by encapsulation of methacryloxypropyl-terminated polydimethylsiloxane (MAT-PDMS) as a UV/sunlight curable agent with urea formaldehyde (MUF).

Microcapsules morphology and particle size distribution were investigated using optical microscopy and scanning electron microscopy. Chemical structure of materials was characterized by Fourier transform infrared (FTIR) analysis. Photopolymerisation rate of MAT-PDMS was also investigated in the presence of benzoin isobutyl ether (BIE) by using FTIR. Also, the elastic modulus and failure behavior of MUF microcapsules were determined through single-capsule compression tests under nanoindenter. At the end, mechanical properties and healing performance of epoxy coating embedded with various numbers of microcapsules were studied.

The results, firstly, revealed that microencapsulation process in the presence of PVP performed successfully. Also, it was observed that most of these microcapsules were spherical and had smooth surfaces. The mean particle size of microcapsules was found to be about 150 μm. The lowest burst points of different particles were found to be in the range of 20 μN and 40 μN. The elastic modulus of microcapsules was measured to be about 13.5 MPa.

The SEM images showed that the crack was filled and healing agent sealed damaged area. Self-healing coating panels were kept at ambient temperature under sunlight for 24 h to have the released MAT-PDMS cured. The self-healing function of coating was evaluated by using salt-spray test. It was found that the self-healed steel panel showed no evidence of corrosion and rust formation when immersed in a 3.5 wt.% NaCl solution for 240 h while the control sample was corroded in this condition.

References:

1. *White SR, Moore JS, Sottos NR, Krull BP, Santa Cruz WA, Gergely RCR. Restoration of Large Damage Volumes in Polymers. Science. 2014;344(6184):620-3*
2. *Wang, Yongjing; Pham, Duc Truong; Ji, Chunqian (2015). "Self-healing composites: A review". Cogent Engineering.*
3. *S.M. Mirabedini, I. Dutil, L. Gauquelin, N. Yan, R.R. Farnood, Self-healing Acrylic Latex Coatings Using Novel Oil-filled Ethyl Cellulose Microcapsules, Submitted to Progress in Organic Coatings, 2014.*
4. *M. Behzadnasab, M. Esfandeh, S. Mirabedini, M. Zohuriaan-Mehr, and R. Farnood, "Preparation and characterization of linseed oil-filled urea–formaldehyde microcapsules and their effect on mechanical properties of an epoxy-based coating," Colloids and Surfaces A: Physicochemical and Engineering Aspects, vol. 457, pp. 16-26, 2014*

Keywords: MAT-PDMS, melamine urea formaldehyde, microencapsulation, self-healing coatings

S.18-3 Self-healing car paints from maize

Gerhard Wenz[1], Gergely Kali[1], Jessika Brunke[2], Carsten Becker-Willinger[2], Nicole Bauer[1]
[1]Saarland University, Saarbrücken, Germany
[2]Leibniz Institute of New Materials, Saarbrücken, Germany

Fine scratches on a new car happen quite often and are always cumbersome. In an application oriented project, we are going to develop a car paint with self-healing properties based on polyrotaxanes. Polyrotaxanes are molecular entities where many rings are threaded onto a polymer backbone hindered by stoppers from sliding off. Covalent cross-linking of threaded rings gives rise to slide-ring gels, which show self-healing properties due to the restricted mobility of the rings in the materials. Self-healing coatings are indeed already on the market for cellular phones, but they still do not meet the tough requirements for car paints. For applications in automobile industry they have to be weather and UV resistant, robust, and finally producible at an acceptable price in large scale.
Our polyrotaxanes are composed from cyclodextrins (CDs) which are cyclic oligomers of glucose readily available in industrial scale through enzymatic degradation of maize starch.[1] Copolymerization of monomers like isoprene and 2,3-dimethyl-1,3-butadiene with styrene in presence of CDs allows the one pot synthesis of polyrotaxanes in water.[2] Homogenous and smooth coatings are obtainable from our polyrotaxanes from solutions in MiBK by casting or spraying. These films can be toughened by conventional cross-linkers, e.g. hexamethylene diisocyanate (HMDI) to form slide ring gels via urethane linkages. Relevant material properties (Martens hardness, glass transition temperature, transmittance) had been investigated and correlated with the molecular structure. Fine scratches at our coatings can be healed by heating them to 100°C for 10 min. Right now, we optimize the synthetic process to allow scale up.

References:

1. *Wenz, G. Angew. Chem. Int. Ed. Engl. 1994, 33, 803.*
2. *Kali, G.; Eisenbarth, H.; Wenz, G. Macromol. Rapid Commun. 2016, 37 (1), 67.*

Keywords: bio-renewable feedstocks, polyurethane chemistry, self-healing coatings

S.19-1 How does automation, digitalization and connectivity changes todays wet grinding and dispersing technologies?

Norbert Kern
Bühler AG, Grinding and Dispersing Technologies, Switzerland

Modern wet grinding processes are designed with smallest beads, high flowrates and maximized power introduction. This means that more and more the machine and plants are operated at the highest possible capacity. To keep this process performance machine controls must be design in new ways. Automation with high process control but also continuous observation of important process parameter building the fundament of sustainable high performance. Smart controls try to put different process parameter in correlation. With these indicators, parameters can be readjusted, wear parts replaced or preventive maintenance scheduled. ConnectivityTo do so machine controls needs to able to up load data in to a cloud to have one single access point. Important is that all this data need to be linked to a manufacturing batch to ensure traceability. This is realized by connecting different controls levels. The majority will be the connectivity to a host system, like plant control systems or SAP for example. But also data exchange from one machine to the other will be important.AutomationTo optimize processes basically we need to go close to the limits of the equipment and materials produced. For this, the process controls need to be able to detect changes of defined parameter. With this information we can identify if the process is getting out of the designed performance window. Either the control can create a message or adjust single or complete set of parameters by itself. DigitalizationAll information regarding process but also regarding product properties has to be digitalized and in a form which can be process with modern control concepts. For this product properties, such as quality (e.g. particle size distribution, viscosity etc.) needs to be formatted to be processed on process optimization algorithms.Another topic is benchmarking, either between processes, machines, locations or even within the industry. This will identify single machines, plants or complete locations with lower production performance. With the exchange of this information all locations (of the same company) can be brought up to the highest possible performance.

S.19-2 24/7 for the perfect formulation

Ellen Reuter[1], Claudia Bramlage[1], Frank Kleinsteinberg[2], Daniel Brünink[3]
[1]Innovation Management, Essen, Germany
[2]Applied Technology, Essen, Germany
[3]Marketing, Evonik Resource Efficiency GmbH, Essen, Germany

Evonik has developed a fully automated testing system for coating formulation, application & characterization on a lab scale together with the HTE supplier Chemspeed.
This HTE equipment is enabling Evonik to support their customers with screening of additives in various formulations and applications to be one step ahead of the competition.
The equipment is two meters high and occupies a 120-square-meter area in an air-conditioned room. It doses the raw materials automatically and, in the first step, formulates them into coatings. In the second step, substrates are coated with the formulated coatings, dried, and transported to the test section, where the properties of the formulations are characterized as soon as the coating formulation is cured. Testing 24/7 is no longer a dream anymore.
The system:

- Operates automatically, provides reproducible results, can be used flexibly,
- formulates coatings from a wide variety of different raw materials,
- characterizes the fluid formulations prior to application,
- coats the panels precisely and dries or cures the coating at elevated temperature or via UV radiation,
- characterizes the finished coatings regarding e.g. film thickness and coloristic properties as well as gloss & haze,
- produces a continuously growing data pool as a basis for future analyses via data mining.

This new system is a very helpful tool e.g. for experiments which are needed to optimize pigmented formulations. As the system allows to vary multiple parameters and optimize them, studies can be carried out regarding dosage of active matters on different pigments as well as studies on combinations of active matters. In addition process parameters like grinding bead size and grinding intensity can be optimized. Broad testing of different raw materials & process parameters allows to identify the best possible solution for the customer. Examples of dispersant studies are shown.
With this broad applicability the new, fully automated high-throughput system enables Evonik to cut down significantly on time spent searching for the optimal formulation or structure of new products - in doing so, Evonik accelerates not only its own pace of innovation, but also that of its customers.

Keywords: data mining, dispersants, high throughput testing, pigment stabilization

S.19-3 Experimental and numerical investigations of atomization processes in paint application using high-speed rotary bells

Oliver Tiedje[1], Qiaoyan Ye[1], Bo Shen[2], Joachim Domnick[2]
[1]Coating Systems and Painting Technology, Fraunhofer Institute for Manufacturing Engineering and Automation IPA, Stuttgart, Germany
[2]Natural Sciences, University of Applied Sciences Esslingen, Esslingen, Germany

The paint material properties and the paint application process parameters determine the properties of the paint film. Here, high-speed rotary bells have increasing importance due to the combination of high qualitiy and improved process efficiency.
A key point for the performance of paint atomizers is the atomization process itself, which, in the case of high-speed rotary bells, is located in a very narrow region close to the bell edge where a thin film is formed that further disintegrates. This process is investigated numerically and experimentally by the present contribution. Special focus is given to the rheology of the paint material and the surface tension. Regarding rheology, the simulation determines the relevant shear and elongation rates, discussing also the appropriate measurement methods for the viscosity (rotational, extensional, visco-elastic). Especially in the case of water-borne materials, the surface tension has to be treated time dependent. The paper describes the relevant time constants.
Methodically, the simulations are based on fluid dynamic models combining the volume of fluid (VOF) method for interface modeling. The simulations cover the paint flow inside the applicator and on the bell surface up to the bell edge as well as the air region in the near bell zone, where ligaments and droplets are generated. The experimental validation is performed by high-speed video analysis of the flow on the bell cup surface and the disintegration zone at the bell edge. High temporal (approx. 32 000 frames per second) and spatial (approx. 5 µm) resolutions are applied. Additionally, laser Doppler anemometry is used to measure the droplet velocities. Also droplet size distributions were determined using laser diffraction and phase Doppler anemometry (PDA), including also a comparison of these two instruments. Another focus will also be drawn to the generation of nano-particular droplets for different paint systems.
The thickness of the film on the bell cup is around 20 µm at the bell edge with resulting shear rates up to 500 000 s-1. Geometrical structures of the bell result in waves of the paint film flow on the bell. With the investigated bell (without serration), the atomization is dominated by turbulence. The droplets first travel in tangential direction before being redirected by the shaping air flow at a distance of approximately 1 mm from the bell edge.
Furthermore, interesting results were found regarding the local correlations between droplet size and droplet velocity inside the spray cone. Close to the bell, the propagation of the big droplets is mainly dominated by the initial atomization process, whereas smaller droplets follow the air flow. Further away from the atomizer all droplets show similar velocities, besides some very small droplets close to the spray cone axis having velocities towards the atomizer. The performed comparison of laser diffraction and PDA show good agreement for several paints, except a highly pigmented white paint material which does not deliver consistent results using PDA.

Keywords: application process, atomization, numerical simulation, rotary bell

S.20-1 Gradient copolymeric coatings via photocontrolled surface segregation

Alessandra Vitale, Roberta Bongiovanni
Politecnico di Torino, Torino, Italy

In this work we describe a facile, single step method for generating gradient polymeric coatings and for independently controlling their surface and bulk properties. Photocured copolymers are obtained by photopolymerization of polyoxyethylene (meth) acrylates and a low polarity comonomer (i.e., siloxane or perfluoropolyether). These copolymers are found to have a compositional gradient, showing simultaneously the bulk properties of the main hydrophilic monomers and a strong hydrophobic character on the surface. Thanks to the spontaneous migration of the apolar chains towards the free surface and to their segregation induced by the photocrosslinking process, coatings having a low surface energy are produced using only a small amount (≤1 wt.%) of the low-surface energy comonomer. The factors governing the surface segregation process, including the light intensity used for the photopolymerization reaction, the film thickness and the monomer structure and functionality, are explicitly investigated.

Keywords: gradient coatings, photopolymerization, surface segregation

S.20-2 Preparation, characterization and application of smart multi-functional polymer coatings

Alexander Münch[1], Petra Uhlmann[1, 2]
[1]Nanostructured Materials, Leibniz-Institut für Polymerforschung Dresden e.V., Dresden, Germany
[2]University of Nebraska-Lincoln, Lincoln, USA

Products with a higher consumer acceptance, an expanded product range and an increased added value can be created by implementation of multi-functional coatings with e.g. easy-to-clean, non-fouling, anti-fog, anti-ice and anti-fingerprint properties. One possibility to realize these requirements is the development of tailored polymer films. To generate those layers, functional polymers specifically designed for the desired function are purposefully synthesized and covalently grafted to the surfaces via one or more attachment groups resulting in stable thin layers. In addition to these specific novel surface functionalities the coating materials have to be applied by using industrial processes which causes particular requirements in terms of coating conditions (velocity, temperature, solvents), reproducibility and sustainability of the films, costs and scalability.

For this propose adjusted co-polymers consisting of a functional unit and particular anchor groups for the covalent attachment on the specific material surface were synthesized by radical polymerization. Zwitterionic components, like sulfobetaine, carboxabetaine or phosphorylcholine, are promising functional units for multi-functional polymer coatings showing the specific properties as mentioned above because of their polarity and high degree of swelling in water. As anchor groups glycidyl and benzophenone groups were selected for thermal as well as UV-induced attachment. The obtained coatings were fabricated with film thicknesses between 3 and 15 nm. Hence, appearance and haptic of the coated products are not influenced by the deposited polymer films.

In this contribution we present a multi-functional coating with simultaneous easy-to-clean, non-fouling as well as anti-fog properties based on co-polymers consisted of zwitterionic phosphorylcholine groups and different anchor units.[1,2,3] The polymer films were fabricated by a fast (<1 min.) grafting procedure at room temperature which is scalable and integrable into fast industrial coating processes, like roll-to-roll or printing techniques. To demonstrate the versatility of the designed polymers, various technical substrates, such as foils, paper and glasses, were functionalized.

Financial support by BMWi is gratefully acknowledged (AiF-IGF 18696 BR; AiF-IGF 18573 BG/1).

References:

1. *A.S. Münch, [...], P. Uhlmann, "Smart functional polymer coatings for paper with anti-fouling properties", J. Mater. Chem. B, in revision.*
2. *A.S. Münch, [...], P. Uhlmann, "Fast preparation of biopassive non-fouling coatings on cellulose", J. Coat.Technol. Res, submitted*
3. *A.S. Münch, [...], P. Uhlmann, ACS Appl. Mater Interfaces, in preparation.*

Keywords: anti-fouling coatings, easy-to-clean coatings, non-fog coatings

S.20-3 How structural order boosts performance – easy-to-clean effects provided by highly defined block copolymers

Lars Hoffman [1], Andrea Huang[2]
[1]Global formulation additives, BASF, Heerenveen, the Netherlands
[2]Global formulation additives, BASF, Hong Kong, SAR China

Highly defined block copolymers are a fascinating class of polymers, as they can combine very different properties into a single polymer molecule. In this paper we present a novel class of block copolymers with perfluorinated side chains in one pol Hong Kong ymer block. Those fluorinated block copolymers are extremely surface active and show excellent compatibility with commercial coating systems.

In this paper we will present data from various analytical measurements (eg. SIMS profiling, surface energy measurements, etc.) on additivated coating surfaces which prove the surface activity of the new polymers leading to extremely low surface energies of the final coating surfaces. The highly structured polymer architecture makes very efficient use of the surface active fluorinated side chains. Compared to conventional fluoro-acrylate additives shorter side chains can be used resulting in more sustainable products.

Two component polyurethane clear coat formulations have been additivated with the novel fluorinated block copolymers. Those surfaces show permanent omniphobic properties which include easy-to clean effects.

Fluorinated block copolymers represent a novel, powerful class of additives for paints and coatings which opens an easy-to-use way of formulating functional coatings with easy-to-clean properties

PARALLEL SESSION 21: Functional coatings

S.21-1 Dolphin inspired compliant coatings for drag reduction of ships – first successful demonstration on ship model

Volkmar Stenzel [1], Claus Schreiner[1], Lars-Uve Schrader[2]
[1]Paint Technology, Fraunhofer IFAM, Bremen, Germany
[2]Computational Fluid Dynamics, Hamburg Ship Model Basin HSVA, Hamburg, Germany

Depending on a ship's speed, surface friction is responsible for 50-80% of the total drag of the ship. The flow condition directly on the bow region is laminar, but it quickly turns into the turbulent regime when water travels along the hull. The transition from laminar to turbulent flow occurs via the development of periodical fluctuations (Tollmien-Schlichting (TS-) waves). The turbulent flow regime causes approx. 10 times higher friction compared to laminar flow. So if it is possible to postpone the transition from laminar to turbulent flow and thus to increase the regime of laminar-flow, a significant reduction of drag of the vessel could be achieved.
The best example for transition delay from nature is the dolphin: dolphins reach an underwater speed of up to 10 knots and the flow around the body is predominantly laminar[i]. From early on it was assumed that the special mechanical properties of the skin of dolphins played a key role here. In the late 1950s a series of experiments was carried out by KRAMER[ii] which indicated that soft coatings inspired by the properties of dolphin skin were able to significantly reduce the drag of surfaces in water. The effect of such coatings is based on the fact that the soft surface postpones the formation and stabilizes the development of TS-waves, meaning that the transition from laminar flow to turbulent flow occurs later. This explanation appeared plausible at the outset, but the effect could not be reproduced in later studies and so doubts about the results of KRAMER arose. Subsequent studies[iii] by *CARPENTER and GARRAD (1986)* have shown, however, that Kramer-type coatings are theoretically able to reduce drag, but the mechanism is considerably more complex than outlined above.
In a recently accomplished research project (FLIPPER) this approach was tackled again in cooperation between Fraunhofer IFAM, Hamburg Ship Model Basin (HSVA), a leading French polymer manufacturer (Arkema) and the Hamburg University of Technology (TUHH). Here the combination of numerical simulations, coating material development and hydrodynamic experiments led to the first successful demonstration of a passive coating that reduces the drag of a ship model significantly. The coating consists of two layers, each with very specific mechanical properties. The inner layer with a thickness of several millimeters resembles the dolphin's blubber and the outer, stiffer layer reflects the dolphin's dermis.
The path from initial numerical simulations of the hydrodynamic phenomena via simulations of material parameters and material development to the final successful demonstration in a water tunnel will be presented.
The authors thank the German Federal Ministry for Economic Affairs and Energy for funding of the project.

References:

1. *Kramer, M.O., The dolphin's secret, Mav. Eng.J., 73(1), 1961, 103-107*
2. *Kramer, M.O., Boundary layer stabilization by distributed damping, J. Aeronaut. Sci., 24(6) 1957, 459-460*
3. *Carpenter, P.W., Garrad, A.D., The hydrodynamic stability of flow over Kramer-type compliant surfaces. Part 1. Tollmien-Schlichting instabilities. J. Fluid Mech. 155, 1985, 465-510*

Keywords: drag reducing coating

S.21-2 Development of repellent coatings based on functionalised silica nano-particles for renewable energy, aerospace and construction.

Geraldine Durand [1, 2], Alan Taylor[1], Marion Bourebrab[1, 3], Angelo La Rosa[1, 2]
[1]Functional Coatings and Resins, TWI Ltd , Cambridge, United Kingdom
[2]Advanced Resins & Coatings Technology Innovation Centre, London South Bank University, London, United Kingdom
[3]School of Engineering, University of Edinburgh, Edinburgh, United Kingdom

Water, dust, oil, inks, ice deposition can generate major cost to industry and are an ongoing cause of failure or loss of performance in many sectors such as Energy, Aerospace and Construction. A solar panel can lose up to 40% of its performance if its surface is soiled whilst a passive anti-icing coating could save £14bn in fuel and 150M tons CO_2 annually from aviation alone. Current mitigation technologies rely on mechanical cleaning such as breaking of the ice. These are expensive, inefficient, unreliable, and damaging to the environment.
TWI ltd has been focusing in developing customised novel passive repellent coating that will delay and prevent any soil formation and soil adhesion without the need for active surface cleaning management. A materials-by-design approach has been adopted where nanoscale silica nanoparticles are being designed to provide reinforcement to conventional resins and also to allow surface roughness. The introduction of multiple functionalities onto the surface of nanoscale silica is the key enabler for the development of a mechanically durable and highly repellent material. These nanoparticles are designed to allow high loading into the resin matrix without affecting its viscosity and homogeneity. The surface chemistry of the nanoparticles can be adjusted to allow the functionalisation to crosslinked with the polymer chain. The full integration of the nanoparticle within the polymer network creates a more durable material without the associated increase in brittleness. Moreover, at the surface, the nanoparticles are expected to both improve wear resistance and, by use of suitable surface chemistry provide significantly enhanced hydrophobicity leading to a new route to durable repellent composite materials.

References:

1. *Makkonen L., Ice adhesion – theory, measurements and countermeasures, J. Adhes. Sci. Technol. 26 (2012), http://dx.doi.org/10.1163/016942411×574583*
2. *Stöber, W.; Fink, A.; Bohn, E. Controlled growth of monodisperse silica spheres in the micron size range. J. Colloid Interface Sci. 1968, 26,62-69*
3. *Wojdyla A.M., Taylor A., Durand G.G., Boyd I W., New assessment criteria for durability evaluation of highly repellent coatings. Wear, 2017, 390–391, 49-60*
4. *Taylor, A.; Durand, G.G.; Alvarez Tirado, M.; Sid, N.; Mycock, S. Functionalisation Method for Metal Oxide Particles. U.S. Patent WO2017/093759A1, 8 June 2017.*
5. *Bourebrab M.A.; Durand, G.G.; Taylor, A. Development of Highly Repellent Silica Particles for Protection of Hemp Shiv Used as Insulation Materials. Materials 2018, 11, 4; doi:10.3390/ma11010004*
6. *Bourebrab M.A.; Sid, N.; La Rosa, A.; Durand, G.G.; Taylor, A. Identification of a threshold of water content during Stöber silica formation–influence on particle characteristics., 2018, Manuscript in preparation.*

Keywords: coatings, hydrophobic, nanoparticles, silica, functionilised

S.21-3 High performance paints by mondo minerals talcs

Veli Kilpeläinen [1], Sander van Loon[2]
[1]R&D, Marketing and Sales, Mondo Minerals B.V. Branch Finland, Kajaani, Finland
[2]R&D, Van Loon Chemical Innovations, Amsterdam, the Netherlands

Durability and resistance requirements of paints and coatings are constantly increasing and at the same time paints need to be healthier for consumers and environmental. Talc as natural, inert and non-toxic material with lamellar and hydrophobic particles is economical and safe choice for many decorative and protective coatings. This presentation demonstrates how MONDO MINERALS talc grades, that are pure and platy, improve the durability and resistance properties of decorative and protective coatings. The wet scrub resistance and surface resistance against different chemicals of decorative paints can be improved when pure, platy talc is used as extender. Also anti-corrosion properties of industrial coatings can be improved and at the same time VOC emissions can be reduced when special MONDO MINERALS talc is used as extender.

Keywords: barrier coatings, corrosion protection, resistance coatings, talc

PARALLEL SESSION 22: New methods for production, characterization and testing

S.22-1 Gamification of Coatings R&D

Rolf Gueller, Carine Marcos
Chemspeed Technologies AG, Switzerland

The coatings industry faces the increased demand for functionalized and innovative coatings combined with time-to-market in this highly competitive field.
Gamification of R&D represents a paradigm shift in coatings development. Virtually unrestricted combination of sequential and parallel processing (Object Oriented Workflow Design) combined with IF/ELSE functions enable automation of intelligent workflows for a tremendous increase in efficiency, flexibility, diversity, standardization and speed in the entire formulation, application and testing process!

S.22-2 Influence of polymer additives in polypropylene on coating adhesion

Katharina Weber, Markus Keuerleber, Volker Wegmann
Fraunhofer IPA, Stuttgart, Germany

Polypropylene-based polymers gain in importance for exterior applications. When it comes to high quality applications, e.g. for the automotive sector, the plastic parts have to be coated. As a result of the required steam jet test (DIN EN ISO 16925:2014-06) in connection with a climatic change test, an increase of adhesion failure is observed in quality tests. It is anticipated that polymer additives have a significant influence on the coating adhesion because of their migration to the substrate surface. However, systematic investigations of this effect are not yet publically available.

We therefore report a detailed study on early indicators for a weakening of the coating adhesion due to the type and critical amount of additives used in the PP-substrate. An analytical approach for the detection and the determination of the surface load of polymer additives such as stabilizers, flow additives and release agents is presented. The aim is to derive knowledge of the critical amount of additive migration to the surface for different additive types. Furthermore, we address the question of how different cleaning and activation processes can reduce the additive coverage on the PP-substrate and thus improve the adhesion. Surface cleaning and pretreatment processes such as dry ice snow cleaning, powerwash, flaming and plasma activation will be compared systematically.

The results to date indicate that the increase of the release agent is most critical for both the coating adhesion and the overall appearance of the coating, due to wetting problems of the base coat. Analytical evidence of the release agent migration up to the primer surface was found via infrared spectroscopy. An analytical GC-MS method for the detection of different additive components with the perspective of quantitative analysis in combination with solvent rinsing of the PP-surface is presented. All analytical results are linked to the corresponding adhesion tests.

Keywords: additive migration, coating adhesion, coating on polypropylene, polymer additives

S.22-3 Phase-stepped deflectometry as a powerful tool to determine the appearance of paints and coatings

Martin Bosma
Corporate innovation Group, Allnex, Wageningen, the Netherlands

Phase-stepped deflectometry (PSD) is a relatively new, powerful optical technique that can be used for the quantitative determination of the surface profile (appearance) of paints and coatings. With the PSD method a computer screen is used to display various horizontal and vertical patterns (phase-stepped sinusoidal fringes) and a camera is used to capture images of these fringes reflected by the surface under study. Analysis of these images allows determination of the local slope at each pixel of the surface, see also the example below. From these local slopes we can obtain the height profile by integration whereas differentiation of the slopes gives the curvature. Appearance of coatings is related to the degree and wavelength of the changes of the local slopes (curvature). Application of FFT filtering allows determination of these characteristics in certain wavelength ranges (e.g. short wave or long wave).
A considerable advantage of PSD is that it is a very affordable technique as it only requires a camera and a computer monitor. In this paper we will describe the underlying measuring principles of PSD. Furthermore, several examples of characterization of surface profiles using this PSD method will be shown. The PSD results will be compared to other methods used in the coatings industry to evaluate coating appearance.

WE43

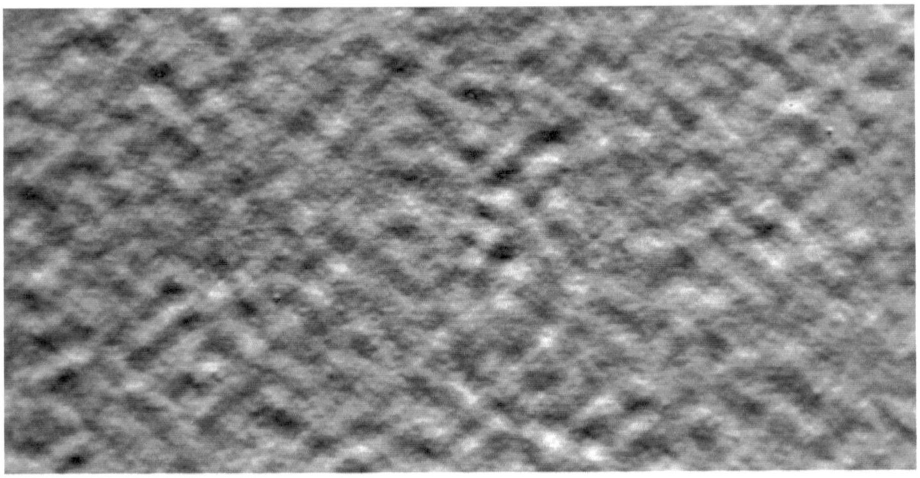

Keywords: appearance, sag and levelling, wavescan

S.23-1 Polypyrrole and zinc-oxide-based anti-corrosive coatings for we43 magnesium alloy

Sasha Omanovic[1], Jun-e Qu[2], Mario Ascencio[1], Li-Min Jiang[3]
[1]Chemical Engineering, McGill University, Montreal, Canada
[2]Faculty of Materials Science and Engineering, Hubei University, Wuhan, China
[3]School of Material Science and Engineering, Nanchang Hangkong University, Nanchang, China

Mg alloys are attractive materials for biodegradable implant applications. However, Mg alloys corrode too fast in the physiological environment, leading to premature decrease in the implant mechanical properties and possible implant failure. In addition, hydrogen gas is formed at quantities that cannot be bio-absorbed by the body, leading to hydrogen gas accumulation and delaying the healing process. Recently, the use of polypyrrole (PPy) coatings for controlling the degradation rate of Mg-based implants has been suggested. Here we present results on the formation of a PPy coating on a WE43 magnesium alloy for the purpose of controlling the corrosion rate of the alloy. Further, zinc has been used in various applications for corrosion protection for many years. Here, we suggesting to use a ZnO-based coating formed on a WE43 magnesium alloy by electrophoresis, also for the purpose of controlling the alloy's corrosion rate.
PPy coatings were successfully synthesized on WE43 Mg alloy. The coatings were homogeneous, had a thickness of about 2 μm and had a fair adhesion to the substrate. Electrochemical impedance, hydrogen evolution and Mg electrolyte concentration results showed that the PPy coating effectively contributes to a decrease in the corrosion rate of WE43 Mg alloy and production of H2 gas, which is favorable for biomedical applications. The corrosion protection of PPy is attributed to a combination of mechanisms including non-specific barrier properties, decreased PPy coating conductivity by the reversible adsorption of dissolved CO^2 species and, misalignment of the substrate and coating Fermi levels caused by uncompensated coating positive charging, eventually inhibiting the transport of Mg2+ ions. Finally, PPy coatings show attractive properties for tailoring the corrosion degradation of Mg alloys for biomedical applications, from its combination of corrosion protection mechanisms to its ability for controlled dopant and drug release; however, optimization of coating properties such as adhesion and coating thickness are still required, as well as the investigation of the coating corrosion behaviour at longer immersion times.
Protective ZnO coatings were successfully formed on a WE43 Mg alloy surface by electrophoretic deposition of ZnO nanoparticles followed by heat treatment. The coatings chemical composition, crystalline structure and surface morphology was investigated by scanning electron microscopy (SEM), X-ray diffraction (XRD) and X-ray photoelectron spectroscopy (XPS). The coating was found to be polycrystalline. The outer layer of the coating was found to contain oxide, carbonates or /and hydroxide of Zn and Mg. Investigation of the corrosion behavior in a phosphate buffered saline solution (PBS) by electrochemical impedance spectroscopy (EIS) combined with corrosion morphology observation by SEM, showed that the synthesized ZnO coatings effectively protected the WE43 surface from corroding, offering the corrosion protection efficiency above 99.9% even after 6h of immersion in PBS.

Keywords: Corrosion Protection, Polypyrrole, WE43 Magnesium alloy, ZnO nanoparticles

S.23-2 Electroactive benzimidazole based polyimides for coating technologies

Humaira Masood Siddiqi [1], Asma Iqbal[1], Naila Khalid[1], O.Ok Park[2]
[1]Department of Chemistry, Quaid-i-Azam University Islamabad Pakistan, Islamabad, Pakistan,
[2]Department of Chemical and biomolecular engineering, Korea Advance Institute of Science and Technology,KAIST, Daegeon, South Korea

Heterocycles based polymeric materials besides their pharmacological properties are also being investigated significantly in a number of optoelectronic applications in form of thin films or uniform coatings such as solar cells, organic light emitting diodes and electrochromic devices. The rigid structure of benzimidazole and intramolecular hydrogen bonding greatly enhances the regularity of polymer backbone leading to improved properties for example high thermal decomposition temperature, glass transition temperatures, mechanical properties and possess a good combination of film-forming capability and properties leading to their potential for high-performance materials. Herein we report synthesis and characterization of a series of blue green fluorescent polyimides bearing electron rich triphenylamine (TPA) in main chain of polymer and benzimidazole as pendant group for this we have synthesized a new diamine monomer (N^1-(4-aminophenyl)-N^1- (4-(2-phenyl-1H-benzo[d] imidazol-1-yl)phenyl)benzene-1,4-diamine) via an inexpensive synthetic route. The diamine monomer was further polymerized with few anhydrides via two step thermal imidization reaction and free standing films were achieved. Introduction of fused TPA-benzimidazole unit imparted some noteworthy structural advantages into synthesized polyimides like (a) Fused TPA-benzimidazole unit improved the charge mobility along the polymer chain (b) the triphenylamine having blocked para position, which rule out the chances of dimerization reactions among polyimide chains (c) Benzimidazole increasing the size of pendant to disrupt chain packing in turn increasing solubility and processability of resulting polyimides. Electrochemical analysis of the prepared polyimides revealed the lower onset oxidation potentials (0.47-0.61 V) and high lying HOMO levels (-4.8 to -5.8 eV) suggesting their use as hole transport materials. The film forming ability, hydrophobicity and processability of these systems make these materials attractive candidates for inkjet printing and different coating technologies i.e. protective garments, safety, aerospace, electronics, automotive industry etc.

Keywords: Benzimidazole, polyimides, optoelectronics, solution processable, spin coating.

S.23-3 Towards enhanced coating adhesion on plastic substrates through UV photografting

Agnès Rannee, Céline Croutxe-Barghorn
Laboratoire de Photochimie et d'Ingénierie Macromoléculaires, Université de Haute Alsace, Mulhouse, France

Roughness and wettability of a surface are two essential parameters to take into consideration for adhesion. In fact, smooth and hydrophobic substrates are considered to be difficult and challenging substrates often presenting defects in adhesion. In a concern of coating deposition onto these substrates such as polyethylene, polypropylene or polyethylene terephthalate, the adhesion phenomenon can be improved through surface grafting. The UV photografting makes it possible to modify the surface properties without pretreatment using a radical process which could be photoinduced. Pioneering works appeared in the late 1950s. Applying a similar process as photoreticulation, they developed a new method to graft macromolecules on polyethylene through a quite complex method. 1 Since improvement of the process was made leading to the proposal of a semi-continuous set-up exhibiting good protection against oxygen inhibition.2, 3 In this work, we first attempt to transpose the efficiency of acrylate monomers employed in photopolymerization to the photografting process. Secondly, the influence of the light dose on the photografting process to control the grafts anchorage and the consequences on adhesion are studied in details. Focus is made on the photografting of isobornyl acrylate (IBOA) on polypropylene and polyethylene terephthalate under different conditions.The photografting efficiency was characterized by ATR-FTIR and Confocal Raman Microscopy. Then, the treated surfaces were observed with SEM to confirm the presence of grafts and their distribution. Finally, wettability modification was assessed with contact angle measurements. Once the presence of grafts on the surface is validated, adherence measurements have been set-up to explain the influence of grafts on adhesion improvement.The control of the grafting of IBOA through UV is highlighted and a correlation between adherence and the grafts distribution is introduced. This study shows new grafting possibilities by changing the nature of grafted monomer.

S.24-1 Hybrid janus colloids: new strategies for design of multifunctional surfaces

Alla Synytska
Department Polymer Interfaces, Leibniz-Institut für Polymerforschung Dresden e.V, Leibniz, Germany

The research focus of our group "Functional Particles & Interfaces" lies on the development and fabrication of multifunctional hybrid materials with tunable interfacial and surface properties on the basis of polymeric-modified colloids with controlled design. In order to address this goal we create a large library of core-shell particles with homogeneous and Janus architectures, with various shapes, sizes, chemical functionality, mechanical and responsive properties and address the fundamental questions about their assembly at interfaces and in dispersions. The obtained knowledge allows us to use these hybrid colloids as building blocks for design of multifunctional biofilm-, and ice-repellent coatings, underwater adhesives, as well as for stabilization / strengthening of interfaces such as emulsions, foams and design of hybrid catalytic active systems for interfacial catalysis. In this talk I will discuss our developments on the synthesis of hybrid polymer-modified Janus colloids and new strategies for the target design of surfaces with controlled icing and biofouling properties.

S.24-2 Ionic liquids and nanomaterials dispersions – innovative additives for coatings and paints

Sebastian Plebst[1], Thomas Schubert[2], Boyan Iliev[2]
[1]Nanomaterials, Heilbronn, Germany
[2]Ionic Liquids, IoLiTec GmbH, Heilbronn, Germany

Ionic Liquids (ILs) are a comparable new class of materials, consisting entirely of ions, which are liquid at unusual low temperatures. The manifold combinations of cations and anions provide a large number of liquid materials with (tunable) unique properties such as high conductivity, high thermal and electrochemical stability, negligible vapor pressure, non-flammability and tenside like character which lead to good solubility of many inorganic materials. Even small amounts of ILs are able to increase conductivity in a variety of polymers, while not influencing their other properties. Some highly surface active ILs can be used to prepare dispersions of nanoparticles such as carbon nanotubes and graphene, which are stable for a long period of time, even if centrifugal forces are applied. Both, the pure ILs and the dispersions of nanomaterials in them show promising performance for many applications, from rheological modifiers, to lubricants, flame retardants and anti-static additives. In such dispersions the excellent properties of both ionic liquids and nanomaterials are combined to have the additional benefit to allow an easy and safe handling. Current state of the art flame retardants are either toxic upon decomposition or of heavy weight, sometimes even both. While common carbon allotropes might not be the most optimal solution, the allotrope graphene shows a quite promising behavior: The hexagonal structure of graphene with its strong C-C-bonds has a decomposition temperature of above 400°C. Its platelet structure, which is acting as an effective barrier for reactive gases, protects coated materials in an efficient way. In addition, graphene contains only the non-toxic element carbon and will decompose completely to CO_2. Furthermore, it is one of the materials with the lowest weight. Due to these special properties of graphene several research groups demonstrated a flame retardant behavior in different polymer matrices. One of the major drawbacks is the high tendency to form agglomerates, finally resulting in a multi layered graphite-type species. Besides the strong tendency to form agglomerates the strong hydrophilic character causes many problems in the often water based coating processes. IOLITEC was able to overcome the issues of agglomeration and negative interaction with most ionic/polar solvents, to obtain an optimal manufacturing process and highest possible performance. In our contribution we will present our most prominent results, in particular that water based and even solvent-free processes can be achieved by applying our techniques. By using them it is possible to achieve coatings, which enhance flame retardant properties of fibers such as aramid, or to achieve a flame retardant character of fibers, like PES/CO (Polyester/Cotton). Besides their use as flame retardant the combination of nanomaterials and ionic liquids allows also the production of high refractive index coatings. Materials such as ZrO2, TiO_2 or other high refractive materials can be further enhanced by adding some high refractive ionic liquids as dispersant.

Keywords: anti static additives, flame retardants, ionic liquid, nanomaterial

S.24-3 Hydroplasticization of acrylic polymers with various acid content: polymer mobility studies

Benjamin Voogt[1], Loes Van de Kamp-Peeters[1], Henk Huinink[1], Bart Erich[1, 2], Jurgen Scheerder[3], Olaf Adan[1, 2]
[1]Applied Physics, Eindhoven University of Technology, the Netherlands
[2]TNO, Eindhoven, 3DSM Coating Resins, Waalwijk, the Netherlands

Plasticization of latex particles by water is of high interest for paint scientists world-wide. Hydroplasticization eliminates the necessity of adding an organic co-solvent to a waterborne latex and would thus ease societal concern on the impact on health and environment. Unfortunately, this also makes coatings more sensitive to environmental conditions. Hydroplasticization of latex particles can be promoted by increasing the number of water-bonding groups in the polymer. Increasing the amount of acid groups in the polymer can increase the amount of water that is physically bound to the polymer, thereby lowering the effective glass-transition temperature Tg of the polymer as described by Flory and Fox.1,2 Therefore the polymer mobility would increase, promoting the interdiffusion of the polymer chains during the latex drying process. Systematically studying the effect of the acid group content on the polymer mobility as function of the water content would thus support the development of fully hydroplasticizable latex.In the current work, we studied polymer mobility by DMA and GARField 1H-NMR relaxometry3 measurements on coatings with methacrylic acid (MAA) contents varying from 2 to 20 wt-%. By conditioning the coatings at various relative humidities (RH), the mobilities of polymer protons are measured with the transversal (T2) relaxation of the polymer protons, which is a direct measure for the mobility of the polymer itself. Supported with Tg measurements using dynamic mechanical analysis (DMA) and water absorption isotherms by gravimetrical determination of the coating water content, relations between polymer mobility and Tg are found. As expected, the Tg decreases with increasing water content and this is most pronounced with 20 wt-% MAA content. Above a certain water content, the Tg no longer decreases. This could indicate that the majority of hydrogen bonds between acid groups are broken by water and additional water in the coating would not significantly affect mechanical properties, and thus Tg, anymore. The signal intensities obtained with 1H-NMR do not vary significantly with water content, which suggests that the signal is dominated by polymer protons. Two T2 relaxations found thus indicate polymer domains with different mobilities, which could represent different polymer Tg's in the coating. Mobility of both fractions increase with increasing water content.So far, polymer mobility measurements using GARField 1H-NMR relaxometry at different humidities gained interesting results. Additional measurements will be done using FTIR, to elucidate water-polymer interactions on a molecular level.

References:

1. *Fox, T. G. & Flory, P. J., J. Appl. Phys. 21, 581–591 (1950)*
2. *Tsavalas, J. G. & Sundberg, D. C., Langmuir 26, 6960–6966 (2010)*
3. *Glover, P. M., Aptaker, P. S., Bowler, J. R., Ciampi, E. & McDonald, P. J., J. Magn. Reson. 139, 90–7 (1999).*

PARALLEL SESSION 25: Functional coatings

S.25-1 Theoretical and experimental investigation of electrical conductivity behavior of epoxy/ graphene coatings

Ra'na Rafiei Hashjin[1], Zahra Ranjbar[2], Hossein Yari[1]
[1]Surface Coating and Novel Technologies, Tehran, Iran
[2]Center of Excellence for Color Science and Technology, Institute for Color Science and Technology, Tehran, Iran

Graphene as a two-dimensional single layer of sp2-bonded carbon atoms is highly conductive due to its extraordinary electron mobility. Incorporating graphene nanoplatelets (GNPs) in polymeric matrices provides coatings or nanocomposites with new characteristics including electrical conductivity. The most significant feature of this type of systems is “percolation threshold”; the volume fraction of GNP at which the transition from an insulator to a conductor occurs. In this research, the correlation between electrical conductivity of epoxy/GNP coating and its GNP content was fully investigated through studying conductivity predicting models and affecting factors, e.g. intrinsic conductivity of components, physical characteristics of GNP, GNP distribution and its interaction with matrix. The coatings (with thickness of 100 μm) were prepared by dispersing GNP (Angstron N002-PDR with electrical conductivity of 473 S/m) in chloroform through a probe sonication followed by adding epoxy resin. UV-visible spectroscopy, Optical microscopy and turbidimetry proved the high stability of GNP dispersion in solvent during the time and their uniform distribution, which are essential for achieving a robust coating with low electrical percolation threshold. SEM helped characterizing the morphology of GNP in solvent and coating. The electrical conductivity values of coatings with various GNP content (0.3, 0.5, 0.7, 1, 1.5, 2, 2.5 wt.%) were obtained as 1.3×[10]^(-8), 3.8×[10]^(-5), [10]^(-3), 8.8×[10]^(-3), 5.8×[10]^(-2), 1.7×[10]^(-1), 9.9×[10]^(-1) S/m, respectively. The electrical percolation threshold appeared at 0.5 wt.% (equal to 0.25 vol.%); which is an appreciable value thanks to selecting proper materials and fabrication procedure. Conductivity behavior of the coatings was evaluated by statistical and thermodynamic based models. Each of them considers some specific factors affecting the percolation threshold. The models were fitted to the data series (electrical conductivity and GNP volume fraction) and their reliability was compared through error estimation (ε). Among the models, the “additive” model was best fitted since it maintained the lowest value. This is due to comprising more crucial factors and comprehensive terms in an almost precise equation. Accordingly, an attempt was made to modify this model. A high coefficient of determination (R^(2)=0.98) and significant decrease (□65%) of value in its modified form confirmed “modified additive” as the most reliable model in predicting electrical conductivity of epoxy/graphene coating. Employing this model reduces costly and time consuming experiments.

References:

1. *K. Kalaitzidou, H. Fukushima, L.T. Drzal, A Route for Polymer Nanocomposites with Engineered Electrical Conductivity and Percolation Threshold, Materials (Basel). 3 (2010) 1089–1103*
2. *Y. Wang, J.W. Shan, G.J. Weng, Percolation threshold and electrical conductivity of graphene-based nanocomposites with filler agglomeration and interfacial tunneling, J. Appl. Phys. (2015)*
3. *M.L. Clingerman, E.H. Weber, J.A. King, K.H. Schulz, Development of an additive equation for predicting the electrical conductivity of carbon-filled composites, J. Appl. Polym. Sci. 88 (2003) 2280–2299.*
4. *F. Lux, Models proposed to explain the electrical conductivity of mixtures made of conductive and insulating materials, J. Mater. Sci. 28 (1993) 285–301.*

Keywords: conductivity models, electrical conductivity, electrical percolation threshold, graphene nanoplatelets

S.25-2 Large-scale spray deposited conductive nanoparticle on bio-based thin films.

Calvin Brett[1, 2, 3], Nitesh Mittal[1, 3], Wiebke Ohm[2], Daniel Söderberg[1, 3], Stephan V. Roth[2, 4]
[1]Department of Mechanics, Royal Institute of Technology (KTH), Stockholm, Sweden
[2]Photon Science, Deutsches Elektronen-Synchrotron (DESY), Hamburg, Germany
[3]Wallenberg Wood Science Center (WWSC), Stockholm, Sweden
[4]Department of Fibre and Polymer Technology, Royal Institute of Technology (KTH), Stockholm, Sweden

Metallic nanoparticle (NP) can be found in a broad spectrum of applications ranging from plasmonics, sensors, photovoltaics to nanocatalysis. The integration of such NP e. g. as electronics on flexible templates as plastic foils, paper and textiles play an essential role. Hence, we applied Airbrush spray deposition - a high volume roll-to-roll manufacturing process - to fabricate nanostructured multilayer thin films containing metal-organic decomposition (MOD) inks on cellulose nanofibril (CNF) thin films[1]. CNF as produced from wood presents a renewable and sustainable bio-polymer which draws increasing attention as template material for flexible, yet disposable electronics. In addition, metal-organic decomposition inks show the ability of functionalization on these templates. One advantage of airbrush spray deposition techniques is that it can be applied to different surface shapes and sizes. By combining airbrush spray deposition with functionalized bio-based hybrid materials as MOD/CNFs we thus aim for scalable, industrially applicable, flexible electronics. In order to fully understand the fabrication kinetics during film growth, we took the advantage of surface sensitive X-ray scattering techniques with high temporal resolution in combination with in situ surface differential reflectance spectroscopy[2].To gain understanding in detail the process and the influences on the material properties we compared the results on soft cellulose thin films to hard silica surfaces. We manipulated the ambient influences by increasing surface temperature which inhibits the aggregation and facilitates the evaporation of carrier solvents and thus changed the nanostructuring[3]. In the industrial perspective, this needs to be fully understood to received best possible material properties with lowest drying times. Our study reveals the very homogeneous CNF thin films with an unprecedented roughness below 2 nm in combination with a tunable wettability within milliseconds. The functionalization with silver nanoparticles from MOD inks resulted in fully conductive thin films analysed in situ employing wide-angle X-ray scattering.

References:

1. *Roth, S. V. J. Phys. Condens. Matter 2016, 28 (403003), 37.*
2. *Schwartzkopf, M.; Santoro, G.; Brett, C. J.; Rothkirch, A.; Polonskyi, O.; Hinz, A.; Metwalli, E.; Yao, Y.; Strunskus, T.; Faupel, F.; Müller-Buschbaum, P.; Roth, S. V. ACS Appl. Mater. Interfaces 2015, 7 (24), 13547–13556.*
3. *Zhang, P.; Santoro, G.; Yu, S.; Vayalil, S. K.; Bommel, S.; Roth, S. V. Langmuir 2016, 32 (17), 4251–4258.*

Keywords: bio-based, conductive, spray coating, thin film

S.25-3 Redox catalysis driving antibacterial coated surfaces for indoor hospital facilities.

Sami Rtimi
Chemistry and Chemical Engineering, Swiss Federal Institute of Technology (EPFL), Lausanne, Switzerland

During the last few decades, the increase of infections by toxic pathogens/biofilms leading to hospital-acquired infections (HAI) has motivated work in the area [1]. More advanced antibacterial films presenting uniform particle distribution, adhesion to flexible non-thermal resistant substrates, mechanical resistance and faster bacterial/biofilm inactivation under light or in the dark are needed due to health concerns [1]. TiO_2 films have been used under light > 387 nm generating highly oxidative radicals as bactericide films for many years [2]. However, its restricted absorption of solar/ visible light and slow bacterial inactivation kinetics has motivated workers to dope TiO_2 with Cu or Ag to shift the absorption to the visible region. This doping also precludes recombination of the photo-generated charges. Stable, adhesive uniform films of TiO_2 sputtered at 130°C for 4 min inactivated bacteria within 40 min. Nevertheless, TiO_2/Cu (Cu≥ 0.1%) films co-sputtered for 2.5 min led to bacterial inactivation < 10 min under actinic light (4mW/cm2) [3-4]. Next, the sputtered Cu for 5-10s (0.01% by weight/ppb levels) on TiO_2-ZrO2 layers on polyester [5] accelerated the kinetics by a factor of 3 with respect to coatings where the Cu was absent.
Since today's focus of interest is to increase the visible absorption of antibacterial surfaces, further work addressed the sputtering of co-sputtered FeOx-TiO_2 for 2 min on thin polyethylene (PE) flexible films. The FeOx-TiO_2 hetero-junctions obtained presented random distribution for both oxides and led to electron injection from the FeOx under visible light to TiO_2 lower lying trapping states inactivating bacteria within 60 min compared to 210min for TiO_2 films by themselves [6]. The most recent work in this area focuses on metal/bimetal films driving bacterial inactivation. Hybrid Cu-Ag (50%/50%) nanoparticle metal-oxide amlgamated films able to generate highly oxidative radicals have been investigated leading to a quasi-instantaneous bacterial inactivation under low intensity actinic light and in the dark [7]. The release of Cu/Ag-ions during bacterial inactivation were below the cytotoxicity levels permitted by the sanitary regulations. Cu and Ag are partially covered with their oxides when exposed to air and the composite Cu-Ag-polyurethane (PU) therefore contains amalgamated CuO-Ag_2O. Due to the difference in redox potentials, Ag will galvanically replace Cu.

References:

1. *A. Kramer, G. Kampf, BMC Infect. Dis. 2006, 6, 130-139*
2. *S. Rtimi, J. Nesic C. Pulgarin, R. Sanjines, M. Bensimon and J. Kiwi, Interf. Focus, 2015, 5 rsfs. 2014.0046*
3. *S. Rtimi, O. Baghriche, C. Pugarin, J-C. Lavanchy, J. Kiwi, Surf. & Coat. Technol. 2013, 232,804-813*
4. *S. Rtimi, C. Pulgarin, R. Sanjines, J. Kiwi, Coatings, 2017,7,20-49*
5. *S. Rtimi, C. Pulgarin, R. Sanjines, V. Nadtochenko, J-C Lavanchy, J. Kiwi, ACS Appl. Mater. & Interf. 2015,7,12832-12839*
6. *S. Rtimi, C. Pulgarin, V. A. Nadtochenko, F. E. Gostev, I. V. Shelaev, J. Kiwi, Scientific Reports, 2016, 6, 30113-30123*
7. *S. Rtimi, R. Sanjines, C. Pulgarin, J. Kiwi, ACS. Appl. Mater. & Interf. 2016, 8, 47-55.*

Keywords: antibacterial coating, magnetron sputtering, redox catalysis, ultrathin films

S.26-1 In-situ probing reactivity at buried interfaces: limits and challenges of local electrochemical impedance mapping and time lapse microscopy

Polina Volovitch[1], Viacheslav Shkirskiy[1], Thomas Sanchez[1], Vinvent Vivier[2]
[1]Institut de Recherches Chimie Paris, Chimie ParisTech, PSL University, Paris, France
[2]Laboratoire Interfaces et Systèmes Electrochimiques, Sorbonne Universités, UPMC, CNRS, Paris, France

The degradation of polymer coated metals is a combination of complex mechanisms involving local processes at buried interfaces. Classical monitoring of the degradation of such a system by electrochemical impedance spectroscopy does not allow to localize the phenomena while more detailed post-mortem morphological and chemical analysis does not provide real time information about the evolution of the interfaces. This explains the increasing interest to localized in-situ reactivity measurements. While *Scanning Kelvin Probe (SKP)* is used to monitor delamination of the polymer during exposure in humid air [1], *Local Electrochemical Impedance Spectroscopy and Mapping (LEIS and LEIM)* seems to be the most attractive tool to monitor the anodic reactivity at metal/polymer interface in immersion conditions and can be conducted in a quantitative way [2, 3]. LEIM seems however be less adopted for cathodic reactivity measurement because of diffusion and not charge transfer limited rate of cathodic oxygen or water reduction. However, the fact that cathodic reactivity generates hydroxide ions and/or gas formation offers a possibility to survey it using *Time Lapse Microscopy (TLM)*.

In the present work first of all we will review the limits, advantages and drawbacks of LEIM and TLM in application to polymer coated metallic substrates (geometry, chemistry and physicochemical characteristics of the polymer and of the substrate). Secondly, we will present our recent development of LEIM and TLM procedures for data acquisition and treatment which permit to access and to quantify in situ the susceptibility of a metal-polymer system to degradation in immersion condition (delamination or blistering) on the example of several systems ranging from model weak polymers to industrial robust systems in presence of soluble Cl-, HCO3- or NH4+ ions. For instance, the effect of solution composition on the kinetics of polymer delamination from an artificial defect can be observed: in Cl containing solutions delamination kinetics is described by a parabolic low typical for diffusion limited processes while in carbonates it evolves linearly with time suggesting surface reaction as a rate limiting step. Finally, the results of *Confocal Raman Mapping* and ex-situ surface analysis of degraded systems will be applied in order to understand the driving forces of local cathodic and anodic reactivity in different systems and to correlate the evolution of the local electrochemical response with the local surface chemistry.

References:

[1] W. Fürbeth, M. Stratmann, Corros. Sci. 43 (2001) 243-254
[2] T. Balusamy, T. Nishimura, J. Anal.Bioanal.Tech. 7 (2016) 1-12
[3] V. Shkirskiy, P. Volovitch, V. Vivier, Electrochim. Acta 235 (2017) 442-452.

Keywords: characterization, degradation, kinetics, metal - polymer interface

S.26-2 Liquid color measurement – a new approach for color testing reveals potential for more productivity in paint production and advanced digital data exchange over the supply chain.

Ralph Jan Wörheide
Orontec GmbH & Co, Dortmund, Germany

When it comes to testing of colour, there a lot of precise instruments, colour correction software packages, automated spray machines etc. in the market. Colorimetry has become a science with a lot of importance to the paint industry, especially if customers with high quality standards should be served.The high scientific level and the precision of the instruments cannot compensate the deficit, that conventional colour testing is still taking place at the very end of the paint production process and takes a lot of time due to the drying process. In addition, it is state of the art to correct colours instead of producing them in the required quality in the first run.Q-Chain LCM is a new technology for measuring colour in liquid condition. The precision is equal to dry methods, but the speed of the test process is much higher. A measurement needs only 3 minutes.The idea behind the Q-Chain philosophy is to consider the whole supply chain regarding quality. It is not enough to test the result of the process, it makes more sense to start reducing material and process variances from the very beginning of the process. To bring the colour measurement into the liquid phase enables participants of the supply chain to stabilize each of these steps and to reduce the variance of the result. Other testing instruments of the Q-Chain product line help to understand process performance of paint systems at the customer paint line.Manufacturers of pigments including effect pigments and tinting concentrates, as well as paint manufacturers are already using Q-Chain LCM successfully.Both elements of Q-Chain LCM – reducing variances and testing time enables paint manufacturers to speed up their production process significantly.In addition, a cloud solution for quality data exchange over the supply chain is introduced. This solution harmonizes quality data in the supply chain with modern IT technologies. It contains IOT (Internet of Things) approaches, which enable users of testing equipment to understand, communicate and analyse data from different tests of paint systems. It is suitable for R&D, quality control and production control. It also helps different companies in the supply chain to exchange data easily and with less effort for the handling of the documents. The cloud solution is working with Q-Chain LCM, but is not limited to this equipment.The presentation will show practical examples of both technologies, the Q-Chain LCM as well as the cloud solution.

S.26-3 Komo exterior joinery performance certification

Robert Daamen[2], Niels Lutke Schipholt[1]
[1]SHR, Wageningen, the Netherlands
[2]SKH, Wageningen, the Netherlands

KOMO testing and certification is recognised as an important instrument for declaration of exterior joinery coatings quality.

The presentation will outline the principles that control the testing and certification of products and gives details of the test methods that are involved and how these relate to product performance in the field.

KOMO is a foundation, founded in 1962. KOMO has been internationally recognised for decades. The KOMO certification is an international quality standard that provides guarantees for both the supply and demand sides of the building industry. It is a highly trusted quality mark that contributes to quality and safety, declares fitness for use and, through third party control, renders trust to the market. This strongly facilitates the market and enables insurability. The KOMO Quality Mark is 100% voluntary and independent. It owes its success purely to intrinsic values that are recognised also by international market players.

SKH is one of these third party certification institutes who performs annual audits at manufacturers of KOMO certified products. SKH, with its extended knowledge of the woodworking industry and relevant industries supplying this industry, focusses on certification in these industries with the intend to promote the use of wood in the construction industry.

Two of the interesting KOMO certification standards for the coating industry are AD 0814 and AD 0817 in which the requirements for primer and topcoat systems on timber are described. The test methods described in these standards are partly the methods as described in the European EN-927 standard. Additionally some of these methods are modelled to answer the performance demands in specific uses in the joinery industry. The so called quick scan method makes these test standards quite unique.

It is even possible to insure the KOMO certified products and processes because the complete production chain, from the wood quality being used to the mounting of the joinery elements at the construction site is controlled by an independent third party.

References:

1. *https://www.komo.nl/ BRL 0801 (2011-12-20). Nationale Beoordelingsrichtlijn voor het KOMO® attest-met-productcertificaat voor houten gevelelementen. Stichting Keuringsbureau Hout, Wageningen, 66 pp.*
2. *AD 0814 (2016-07-05). ASSESSMENT DIRECTIVE FOR THE KOMO® PRODUCT CERTIFICATE FOR FILMFORMING COATINGS FOR APPLICATION ON TIMBER. Stichting Keuringsbureau Hout, Wageningen, 21 pp.*
3. *AD 0817 (2008-06-10). ASSESSMENT DIRECTIVE FOR THE KOMO® PRODUCT CERTIFICATE FOR FILM FORMING MID- AND TOP COAT SYSTEMS ON TIMBER. Stichting Keuringsbureau Hout, Wageningen, 18 pp.*

Keywords: Certification, exterior, joinery, komo

S.27-1 Graphene coating for heat removal, sound isolation and energy harvesting

Gorkem Memisoglu[1], Burhan Gulbahar[2], Omer Gokce Kuyucu[1], Onder Aydogan[1]
[1]Vestel Electronics, Manisa, Istanbul, Turkey
[2]Applied Research Center of Technology Product, Ozyegin University, Istanbul, Turkey

Graphene is promising for important applications in surface technologies with high ratio of area to height, ultra-low weight, durability and atomic thickness [1, 2]. Graphene aerogel is recently proposed for various applications including supercapacitors, filters and ultra-light material forms with significantly large number of microscale hole formations [3]. In this study, the outstanding properties of graphene are exploited for outdoor air conditioning (AC) units as material coatings to minimize energy losses due to noise, heat and vibration [4].
The coating design includes stacks of graphene layers which generate electricity by utilizing piezoelectric material properties while absorbing heat exploiting thermoelectric properties. The special design allows combining heterogeneous structures to provide piezoelectric, thermoelectric and sound absorbing properties. Therefore, a multi-layered blocking structure is provided for coupling of wind force, heat and acoustic waves travelling in the air with piezoelectric and thermoelectric transduction mechanisms. The multilayer stack has a height of less than five millimeter while preferably including more than thousands of graphene layers with nanometer accuracy layer spacing's allowing special heterostructured design. The design guides heated air of AC units while air force, heat energy and the acoustic vibrations are all absorbed gradually as they flow through the multi-layered structure. In an intercalated graphite structure, the distance between each single layer graphene is 0.575 nm in average. This means, in a height of 1 mm graphite structure, two million single layer graphene absorbers exist. This cannot be achieved currently available macro-scale or micro-scale piezo resistors, heat absorbers or isolation layers preventing the acoustic noise to radiate.
Energy harvesting device promises applications to be utilized on the walls or surfaces of buildings, indoor air conditioning units, outdoor air conditioning units, vehicles, cars, car motors. It operates in any temperature or challenging environmental conditions with better efficiency in terms of heat and acoustic force absorption.

References:

1. *Geim, A. K. & Novoselov, K. S. The rise of graphene. Nature Mater. 6, 183–191 (2007).*
2. *Geim, A. K. Graphene: status and prospects. Science 324, 1530–1534 (2009).*
3. *ZHANG, Qiangqiang, et al. 3D printing of graphene aerogels. Small, 2016, 12.13: 1702-1708.*
4. *B. Gulbahar, G. Memisoglu, "Energy harvesting unit, outdoor air conditioning unit comprising an energy harvesting unit and method with multi-layered absorbing graphene infrastructure," EP16205558.6-1556, EPO Patent Application, December 2016.*
5. *Geim, A. K. & Novoselov, K. S. The rise of graphene. Nature Mater. 6, 183–191(2007).*
6. *Geim, A. K. Graphene: status and prospects. Science 324, 1530–1534 (2009).*
7. *ZHANG, Qiangqiang, et al. 3D printing of graphene aerogels. Small, 2016, 12.13: 1702-1708.*
8. *B. Gulbahar, G. Memisoglu, "Energy harvesting unit, outdoor air conditioning unit comprising an energy harvesting unit and method with multi-layered absorbing graphene infrastructure," EP16205558.6-1556, EPO Patent Application, December 2016.*

Keywords: energy harvesting, graphene coatings, heat removal, sound Isolation

S.27-2 Smart electropainting with graphene

Gorkem Memisoglu[2], Burhan Gulbahar[1], Omer Gokce Kuyucu[2], Onder Aydogan[2]
[1]Applied Center of Technological Products, Ozyegin University, Istanbul, Turkey
[2]Vestel Electronics, Manisa, Turkey

Graphene with significant features promise important applications in printable and laminated structures. Graphene inks provide skin laminated electronics as next generation biomedical sensor network components while graphene empowered glues and dyes are also studied in literature [1]. Laminated heterostructures of two-dimensional materials such as graphene, black phosphorus and transition metal dichalcogenides (TMDs) are utilized for optical modulation of electromagnetic radiation in IR and visible wavebands [2]. In this work, printable graphene ink is used for graphene painting onto the wall by using methods such as ink-jet printing or spray-coating [3]. The painting with a special laminated heterostructure provides smart coating capability denoted by smart electropainting where optoelectronic properties is changed as desired to give the surface a variety of different functions.

Electropainting process utilize graphene layer to electrically tune several different modes of operation. The laminated heterostructure comprises graphene layers, dielectric-metal compound layer such as doped silicon or aluminium and a control unit configured to tune the relative proportions in which the heterostructure reflects and absorbs incident radiation by changing the optoelectronic properties at specific wavelengths. It is capable of transmitting data by optical modulation of incident radiation, harvesting energy from incident radiation or reflecting incident radiation. Smart electropainting process includes a user controllable energy unit with different modes of operation: non-effective mode, cinema/dark mode (patterned area totally absorbing visible light), data transmission mode (dominant modulation and IR/VIS spectra reflection) and heater mode (IR radiation totally reflected).

The proposed architecture promises application areas such as surfaces of walls, domestic appliances or electronic devices. Data transmission mode of operation is used to communicate status information from the domestic appliance or electronic device, either by modulation of a data signal in the visible or infrared waveband, or in the form of a visual display. Energy harvesting mode is used to supply power to the appliance or device, thereby reducing its overall power consumption. The device is tuned to reflect substantially all of the incident radiation in IR band for heating and insulation applications by preventing the leakage of thermal radiation through a surface on which the laminated heterostructure is applied. It is possible to control the operation of the apparatus remotely by utilizing the control unit with wireless communication on a wall, ceiling, floor or roof of a building.

References:

[1] X. Han et al., "Scalable, printable, surfactant-free graphene ink directly from graphite", Nanotechnology, vol. 24, no. 20, p. 205304, 2013.

[2] Z. Sun, A. Martinez, F. Wang, "Optical modulators with two-dimensional layered materials", Nature Photonics, vol. 10, no. 4, pp. 227--238, 2016.

[3] B. Gulbahar, G. Memisoglu, "Laminated heterostructure device, apparatus comprising the same, and method of operating a laminated heterostructure device", EPO Patent Application, EP17172204.4-1903, May 2017.

Keywords: electropainting, graphene coatings, patterning, tuneable colour

S.27-3 Laser-assisted curing of coatings - substitution of oven-heating

Christian Schmitz, Bernd Strehmel
Institute for Coatings and Surface Chemistry, Niederrhein University of Applied Sciences, Krefeld, Germany

Thermal processes for curing of coatings are irreplaceable in most applications with high demand on mechanical and chemical resistance. Oven processes typically provide an environment temperature in the range between 80 - 180°C resulting in sufficient thermal energy needed to heat up both the coating as well as the substrate to initiate chemical and physical processes. It should be attended that the curing time relates to the progress of the temperature curve and the time needed maintaining at this object
temperature to complete conversion of the reactive resin. Alternatively, the use of laser processes for thermal activation should reduce energy consumption, space and processing time. The laser emission results in an increase of the local temperature of the coating for a short duration depending on both the dwell time and the laser power. During the time of laser exposure, the chemical curing and physical processes such as drying or melting and film formation of a powder coating can be induced by heating
particular the coating and not the substrate with its surrounding. The fact that the process would be accessible for laser machining or not mainly depends on the minimum temperature and the time needed for completion of the thermal procedure. Nevertheless, it has been known that the use of near infrared lasers in chemical processes can easily result in high conversion of radiant energy into heat by the use of appropriate absorbers. Thus, it could fit to the requirements regarding a sufficient curing process. Due to a high probability of internal conversion compared to luminescence as a deactivation process of many absorbers in the near infrared range, the temperature of the coating can be increased in a short time frame. The curing of different coating systems such as unsaturated polyester or acrylate resins
combined with a thermal initiator, blocked isocyanates with hydroxy-functional resins, and melamine-formaldehyde resin based coatings were tested with respect to their compatibility for laser-induced curing. Heptamethinecyanines represent one possible group of absorbers for lasers emitting at 808 nm and 980 nm because of their outstanding solubility in coatings compared to rylenes. Their use results in a cured film with a width related to the diameter of the focus of the laser spot by moving the substrate under continuous wave laser exposure. A larger area of cured material was achieved by using a laser setup, which generates a line-focused laser being moved across the coating surface. The curing of powder coating represents a special application for laser-induced film formation where a process speed of around 10 mm·s-1 was achieved.

Keywords: laser-induced curing, NIR diode laser, powder coatings, thermal reaction

S.28-1 Organic coating corrosion protection as seen from the metals

Herman Terryn
SURF Electrochemical and Surface Engineering, Vrije Universiteit Brussel, Brussels, Belgium

In the advanced materials industry, there is a clear trend towards more sustainable concepts, and this is certainly also the case for metals. They take a prominent position in the materials market thanks to their multifunctionality and structural and mechanical properties. However, metals are prone to corrosion, and therefore often organic coatings are applied to extend their lifetime. The search for organic coated metals with higher multifunctionality and extended lifetime goes together with the search for more sustainable material concepts in the context of the REACH regulation.
Nowadays, materials are developed for providing components with a lifetime ranging from 10 to 25 years. The lifetime of a material highly depends on its exposure to the environment inducing ageing processes. In the current state of the art, lifetime and ageing assessment are performed by means of experiments, combining accelerated and field testing. The limitation of the first is that the conditions of the accelerated tests are not representative for the real environmental conditions and that there is no proven relationship between accelerated laboratory tests and field performance data. The limitation of the second is that it takes several years (5 to 10 years).
The only way to realise a substantial decrease in the development time of new materials is to introduce modelling in the design cycle. What is needed to realize a breakthrough in this field is a tool that can predict quantitatively and dynamically the corrosion behaviour of (organic coated) metals. The long term scientific motivation of our research is to build a knowledge and technology platform to enable the prediction of durability behaviour and the estimation of lifetime of (organic coated) metals under long-term environmental ageing and corrosion conditions. This requires advanced research because corrosion of a (organic coated) metal is the result of an intense interplay between several physical phenomena that need to be characterized in real conditions and modelled. Within VUB and TUDelft research we try to focus on both aspects bringing in new advanced combined electrochemical & in situ surface analysis, advanced finite element electrochemical modelling and more recently quantum chemical modelling. During the lecture an overview will be given on our state of the art in the different domains, emphasizing some success but also bringing bottlenecks when it really comes to predictions of corrosion.

Acknowledgements
all VUB-SURF and TUDELFT-CTE colleagues for the excellent collaborations.

Keywords: Corrosion, interface, organic coatings

S.28-2 Mechanism of pigments migration in polymer coatings

Malgorzata Kopec[1], Brenda Rossenaar[1], Kees van Leerdam[1], Tony Davies[1], Stuart Lyon[2]
[1]AkzoNobel Chemicals B.V., Deventer, the Netherlands,
[2]School of Materials, University of Manchester, Manchester, United Kingdom

Anticorrosive pigments provide polymer coatings with active corrosion protection. Effective pigments exhibit limited solubility in water which prevents excessive leaching and results in the long-term release behavior providing high level of anti-corrosion performance during the entire service lifetime. There is a growing interest to search for a replacement of high-performing but toxic chromate-based pigments (the most efficient active corrosion inhibitors). One of the main obstacles in design of novel systems is the lack of understanding of fundamental transport phenomena of pigments in the coatings [1]. The through-coating migration of sparingly soluble species originating from pigments is related to the general problem of ion transport in polymers. It is assumed that ions can penetrate into a polymer matrix through conductive pathways, whose formation is assumed to be correlated with inhomogeneity in the polymer structure; however the detailed mechanism is not fully understood [2]. Additionally, presence of certain additives such as anticorrosive pigments and non-functional fillers can further provide alternative migration pathways [3].
This work aims to investigate the migration mechanism of chromate ions in model polymer coatings and the influence of filler particles on that process. Free-standing coating samples were exposed in the sodium chromate solution under different conditions. SEM-EDX (scanning electron microscopy-energy dispersive X-ray spectroscopy) and STEM (scanning transmission electron microscopy)-EDX were used to track chromium location and analyse changes in the coating structure after exposition. ICP-MS (inductively coupled plasma mass spectrometry) was used to evaluate the overall concentration of chromium in the samples. A model epoxy-amine coating was used in this investigation and the influence of different fillers (TiO_2 and BaSO4) and their volume concentration on the chromate ingress into the samples was examined. It was shown that BaSO4 particles can introduce migration pathways for chromate transport in the coating at specific conditions while the influence of TiO_2 on the migration process was rather limited. Moreover, the results suggest that ion exchange is a significant part of the migration mechanism in the samples with filler particles.

Acknowledgements: Members of the ECG-MAS AkzoNobel are kindly acknowledged for their contribution in the analytical work related to this project. We would like to thank Lennaert Klerk for the contribution in the project and Peter Visser for help with the samples preparation. This work was supported by the European Union's Horizon 2020 research and innovation programme under the Marie Sklodowska-Curie grant agreement No 706908.

References:

1. *S.B. Lyon, R. Bingham, D.J. Mills, Advances in corrosion protection by organic coatings: What we know and what we would like to know, Progress in Organic Coatings 102 (2017) 2-7.*
2. *S. S. Jamali, T. Suesse, S. Jamali, D. J. Mills, Y. Zhao, Mechanism of ionic conduction in multi-layer polymeric films studied via electrochemical measurement and theoretical modelling, Progress in Organic Coatings 108 (2017) 68-74.*
3. *S. Sellaiyan, A.E. Hughes, S.V. Smith, A. Uedono, J. Sullivan, S. Buckman, Leaching properties of chromate-containing epoxy films using radiotracers, PALS and SEM, Progress in Organic Coatings 77 (2014) 257-267.*

Keywords: corrosion protection, ionic transport, pigment migration, polymer coatings

S.28-3 Reciprocity in weathering exposure and the kinetics of property degradation

Stuart Croll
Coatings and Polymeric Materials, North Dakota State University, Fargo, USA

Accelerated exposure is produced via greater irradiation, higher temperatures, more moisture, and rapid cycling. The hope is the kinetics of degradation scale in a simple way and accelerated testing can simulate the more moderate natural world and so one can estimate service life. If "reciprocity" applies, it means that degradation is strictly, and simply, proportional to the integral product of radiation intensity and time, so linking accelerated exposure and the real world can be done. Reciprocity is very appealing and basic, and may be accurate in some applications [1], but when monitoring the properties of coatings, it becomes confused with other factors and difficult to apply. This is an attempt to clarify how the concept is useful and how it becomes obscured.

Reciprocity has been observed in spectroscopic investigations of photodegradation, but there are problems when the coating composition has several ingredients that change differently with exposure, there are temperature effects, or the testing cycle is too short compared to the viscoelastic properties of the polymer. Needs that engage more than one process or one aspect of the degradation of a coating, e.g. corrosion protection, will not exhibit reciprocity because not all the processes react in step to the environment, e.g. moisture diffusion. However, starting with simple kinetics, the applicability of reciprocity can be explored. Necessary considerations obscure the underlying reciprocity, even in properties that follow the average of degradation, e.g. in gloss loss, yellowing or simple mass loss. Even gloss [2] depends on surface structure in a way that is not linearly related to photo-oxidation, so reciprocity is obscured. A property requirement may depend on the greatest extent of degradation that has occurred rather than the average, e.g. mechanical durability or corrosion protection, when extreme value statistics [3] show that we cannot expect to see reciprocity clearly even though the underlying degradation chemical kinetics are no different.

In trying to connect results from accelerated weathering to those from natural exposure we find that mechanisms may have proceeded to a different extent and thus produce a different failure mode. In that case, not only would any accurate descriptive model be complex, but there is little chance that it would be predictive, and thus not cost effective. Reciprocity implies very simple kinetics for photodegradation, but provides a basic approximation that is natural and easy to understand and thus is useful, but it is not predictive for practical applications.

References:

1. *J. W. Martin, J. W. Chin, T. Nguyen, "Reciprocity law experiments in polymeric photodegradation: a critical review," Prog. Org. Coatings, 47, 292 – 311 (2003)*
2. *J. T. Bendler, S. F. Feldman, H. Hatti, S. Y. Hobbs, "Approximate Model of Diffuse Reflectance From Rough Polymer Surfaces." J. Appl. Phys., 83(2), 998–1004 (1998)*
3. *J. – Y. Fortin, M. Clusel, "Applications of extreme value statistics in Physics," J. Phys. A: Math. Theor. 48 (2015) 183001*

Keywords: accelerated weathering, kinetics, reciprocity

S.29-1 VAE technology for waterborne decorative paints with high pH

Lenine De Sousa Gomes
Wacker Chemie AG, Burghausen, Germany

Waterborne technology has become the standard for modern decorative paints. More than 80% of all architectural coatings worldwide are produced using aqueous formulations. Among waterborne paints, high-pH formulations have only played a minor role in the past. However, recent changes in biocide regulations have sparked new interest, as high-pH environments are regarded as possible alternatives for producing biocide-free formulations.
The correct choice of raw materials plays a key role in the formulation of high-performance decorative waterborne coatings. Additives and binders account for the properties and functionalities of the resulting paints. Organic polymer binders can be tailored to allow the paint to be adjusted to suit specific needs.
A paint can be used for many different purposes. Organic-polymer-based binders can be easily adjusted to produce the properties required. For example, an organic polymer with a strong stabilization system is able to withstand highly alkaline environments, such as those found in paints with high pH values.
The aim of this work was to develop indoor matt paints based on vinyl acetate-ethylene (VAE) binders. Two types of paint formulations were considered in order to achieve at least 10.5 units of pH in paints, one using a mineral co-binder (waterglass) and another using a pH adjuster. The waterglass considered in this case was based on potassium silicate and the pH adjuster on potassium siliconate. Both recipes were formulated without coalescents.
The main focus of this study was to optimize storage stability and workability of the paint formulations, since they are considered as challenging properties when organic binders are exposed to highly alkaline environments. VAE-based paints were shown to have the required storage stability and workability when formulated with 1 % w/w of a Polydimethylsiloxane and 0.5 % w/w of a Siliconate on total formulation. Furthermore, it could be shown that paints based on VAE binders also outperform conventional commercial styrene acrylic in terms of hiding power.

Keywords: high pH paint, silicate, siliconate, VAE

S.29-2 Silica-graphene aerogels as the functional component of coating materials

Katarzyna Suwala, Maria Zielecka, Marek Panasiuk
Industrial Chemistry Research Institute, Warsaw, Poland

Preparation and characteristics of a new type of silica-graphene aerogel, combining the physical and chemical properties of silica aerogel and grapheme will be presented. The influence of the graphene oxide content on the physical properties of silica aerogels was investigated. Silica-graphene aerogels were obtained by using the sol-gel process and eliminating drying under supercritical conditions. This processing way allows a direct control of the carbon phase and a good dispersion thereof within the silica matrix. The obtained samples were prepared using different contents of graphene oxide, i.e. from 0.01 to 0.2 with respect to TEOS. The effect of the graphene oxide will be presented based on the results of scanning electron micrographs (SEM), Fourier-transform infrared spectroscopy (FTIR), X-ray diffraction (XRD) and X-ray photoelectron spectroscopy (XPS). The resulting silica-graphene aerogels have very high surface area (593 – 706 m^2/g), low bulk density (0.037 – 0.116 g/cm^2), low thermal conductivity (0.024 – 0.029 W/mK), hydrophobic properties and significantly lower dust emissions.
In conclusion, with the use of sol-gel technology, high-quality silica-graphene aerogels can be obtained with well-dispersed graphene oxide in a silica matrix. The resulting silica-graphene aerogels can be applied the functional component for the production of a variety of materials, giving them specific properties, for example, they can be used as carriers modifiers, and the same as nanofillers for polymer composites. In addition aerogels can be used as an additive to paint, to improve their insulating and acoustic properties.

Keywords: silica-graphene aerogel, sol-gel method

S.29-3 Optimisation of mineral blend to provide light scattering without titanium dioxide in decorative matt paint.

Amandine Briand[1], Anabelle Legrix-Elton[1], Petra Fritzen[2], Emmanuel Bertin[3], Helen Dollani[1], Lieven Verstuyft[4], Catherine Maillard[3], Andrew Curtis[1]
[1]S&T, Imerys, St Austell, United Kingdom,
[2]S&T, Imerys, Rheinberg, Germany
[3]S&T, Imerys, Toulouse, France
[4]S&T, Imerys, Gent, Belgium

Titanium dioxide (TiO_2) is the most commonly used white pigment in the paint industry. Its high refractive index relative to other paint components confers high light scattering pigment properties, providing both wet and dry opacity and contributing positively to the colour of the paint. Due to on-going pressure on the TiO_2 market and the fact it faces regulatory ruling in Europe, paint producers have a constant interest in finding alternative ways to formulate paints with minimum amount of TiO_2. Successful optimisation of formulations allows them to reduce costs and improve the environmental footprint by using the right combination of extenders. This paper reports on the use of inorganic minerals to optimise light scattering in paints without TiO_2.

European paper coating formulations use TiO_2 infrequently, instead mineral blends have been extensively studied to improve opacity and whiteness. The difference in bulk refractive index between any mineral and surrounding medium is too small to directly contribute to scattering, but in such high Pigment Volume Concentration (PVC) systems, the right packing structure of minerals provides porosity and light scattering. In a way, paper coatings could be considered as very high PVC paints without TiO_2 but with a difference in the gloss requirement. Unlike paper coatings, decorative high PVC paints tend to have a low sheen.

Here, we used a Design of Experiment (DoE) approach to study the impact of various mineral factors on the light scattering and size of pores present in the paint film. DoE is a useful tool to evaluate the impact of factors, in our case mineral shape, particle size and internal/external porosity, on responses, i.e. paint properties, with a minimum of experiments. We have used a mixture design to combine several inorganic minerals and we report how this approach has allowed us to understand how mineral properties affect scattering and opacity performance. We can therefore find optimum combinations of minerals to achieve the highest light scattering efficiency in paint without TiO_2 whilst also checking the effect on gloss and other parameters such as viscosity, whiteness, stain and scrub resistance in the whole experimental design. Additionally, further analyses of the pore size and volume structure were carried out in a similar way to paper coating structure.

Keywords: matt paint, porosity, scattering, titanium dioxide

S.30-1 Impact of TiO_2 particle design on weather resistance of high quality paints: a comparison between exposure methods

Johan Rommens[1], Annemie Verhaege[1], Michael Diebold[2]
[1]Titanium Technologies, Chemours, Kallo, Belgium
[2]Titanium Technologies, Chemours, Wilmington, USA

Titanium dioxide pigments play an important role in protecting white and light colored paints against the influences of weathering. Careful design of the pigment enables optimal protection and guarantees maximum lifetime of the coating. Results from our most recent exposure study underline the importance of choosing the right measurement technique to study weathering influences for the reliable development of high quality and durable coatings. This paper is an extension of the work presented in 2016 and includes most recent data from a multi-year exposure study as well as an updated view on the underlying mechanisms that can lead to coating degradation due to weathering.

References:

1. *Coatings World - September 2017, P142-145*

Keywords: accelerated weathering, titanium dioxide, weather resistance

S.30-2 Morphology-control on the nanoscale: recent developments in the encapsulation of metallic effect pigments

Oliver Bedford, Stephanie Hiltl, Oliver Struck
Research & Development, Eckart GmbH, Hartenstein, Germany

Today, corrosion protection of metallic effect pigments (aluminium or brass) through encapsulation with silica (SiO_2) is state of the art. Especially premium products for powder coatings or water-based industrial and automotive applications are broadly available. Although SiO_2-treated pigments offer outstanding properties regarding stability, optics, hiding power and ease of use, this is not enough for tomorrow's markets with increasing and more diverse requirements.
In our work, we show how the versatility of the underlying Sol-Gel chemistry allows meeting these requirements. After introducing the general silane-based encapsulation, we focus on new concepts and present some selected applications. Using the classical Sol-Gel process as starting point, one aspect of the current research is to increase the compatibility of the silica phase to other organic or inorganic phases. This multi-phase approach effectively leads to improved and new properties, like e.g. corrosion protection, colour, or electrical conductivity. It is crucial for the application performance to control the morphology of the nano-sized domains; unwanted side effects include agglomeration, and loss of orientation of the metallic pigments.
The interactions across the inorganic-inorganic or inorganic-organic interfaces are tailored either by organo-silanes, by modification of the non-silica phase, and/or by other metal oxides (MO_x).
Following this general principle, it is possible to prepare organic-inorganic hybrid and MO_x encapsulations, which provide strongly improved corrosion resistance in aqueous lacquers, i.e. reduced gassing even under intensified conditions (high pH, presence of corrosive ions).
By the same principle, organic and inorganic pigments are incorporated in the SiO_2-matrix leading to attractively coloured metallic effects.
As an example for functional applications, we show how the interface design allows the preparation of conductive pigments with enhanced performance.
By including chemistries beyond silica, the Sol-Gel toolbox for stabilization of metallic effect pigments is significantly expanded and able to fulfil the constantly growing requirements regarding aesthetic appearance and functional performance.

Keywords: aqueous coatings, conductive pigments, effect pigments, sol-gel process

S.30-3 Formaldehyde-free fluorescent pigments with unique chemical and light stability!

Rami Ismael
ARALON COLOR GmbH, Heiligenroth, Germany

Functional pigments / daylight fluorescent pigments

Production and use of formaldehyde-free daylight fluorescent pigments

For about a decade, fluorescent colors are an integral part of sports goods. Whether wetsuits, buoyancy aids, ski, bicycle helmets, balls of all kinds, sports shoes or goalkeeper gloves, all luminous colors have been introduced. Functionality plays an important role in addition to the appearance. Every sports has its own properties. This will be briefly discussed using the example of goalkeeper gloves here.
Goalkeeper Gloves are a technical masterpiece. Whether a team will win the FIFA World or European Championship, depends not least on whether the goalkeeper can hold the ball safely. Here, the latex foam plays an important role.

Keywords: fluorescent pigments

S.31-1 Substance to substance bonded metal-plastic joints by the use of latent reactive powder coatings as adhesive - material and technology development

Michaela Gedan-Smolka[1], Ines Kühnert[1], Peter Scholz[2], Didier Garray[3]
[1]Leibniz-Institut für Polymerforschung Dresden, Dresden, Germany
[2]Fraunhofer IWU, Chemnitz, Germany
[3]Sirris – The Collective Centre of the Belgian Technological Industry, Liege, Belgium

Caused by the demand of lowering the fuel consumption and the aimed e-mobility the automotive and transportation industry enhance their efforts of weight reduction. For that reason structures in multi-material design are used in increasing amounts. Large differences of thermal expansion coefficients make it difficult to join metals and polymers. So the connection of incompatible materials is mainly performed by form or force fit. The steadily increasing use of those material combinations and additional requirements regarding with corrosion protection and impermeability make it necessary to change the technology from mechanical joining to adhesive bonding. Therefore usually a further technology step is needed whereas the glue is applied, followed by the assembling. Our goal was to develop powder coating material with an additional well defined latent adhesive function in combination with a technology which enables highly efficient production lines for metal-plastic hybrids.
As found in the IPF earlier, in the presence of Zn-II-acetylacetonate the reaction between uretdione cross-linkers and OH-group polyester results in a two-step curing mechanism. At lower crosslinking temperatures up to 150°C a polyallophanate network is formed selectively. At temperatures ≥ 190°C the polyallophanate network can be transferred into polyurethane structures very fast and completely. Already in the polyallophanate state the powder coating material resists high deformation and complex forming operations and therefore the powder can be applied to a metal substrate, in our case aluminum, as a coil or cut board application. Based on this two-step curing mechanism various low temperature curing and highly flexible powder coating formulations were developed in this study. Furthermore, the shaping and assembling behavior of an adhesive polymer-aluminum composite was examined whereby the polyallophanate layer acts as latent reactive adhesive and high quality surface finish simultaneously.
To generate the metal-polymer hybrid, a thermoplastic polyurethane (TPU) was over-molded onto the pre-coated aluminum substrate by injection molding. In this way aluminum-polymer joints based on flexible polymers can be kept. On top of the TPU a polyamide 6 layer can be over-molded to generate a stiff composite structure. The TPU in between the pre-coated metal and the polymer top layer acts as stress and strain compensation which can result from thermal expansion as well as external deformation. A covalent bonding between the powder coating film and the thermoplastic polyurethane generates substantially higher adhesive strength in contrast to physical adhesion for the hybrid composite. In addition, the described aluminum-polymer hybrids can be fabricated in a shortened process chain. The prefinished coil or cut board material can be formed (e.g. deep drawing) first and inserted to the over-molding procedure afterwards or in the same machine. In this manner an efficient coating process with less cleaning steps and a high versatility in the follow up composite fabrication can be achieved.

Keywords: injection overmolding, latent reactive adhesive, metal-plastic joints, uretdione powder coating

S.31-2 Smart stratifying powder coatings

Marc Entenmann[1], Michaela Gedan-Smolka[2], Heinz Greisiger[1]
[1]Fraunhofer Institute for Manufacturing Engineering and Automation, Stuttgart
[2]Leibniz Institute of Polymer Research Dresden, Germany, Dresden, Germany

Multilayer coatings are state of the art for today's high quality automotive and industrial protective coatings systems. As many as three to four separate layers are nowadays used for the corrosion protection and optical properties of metals. Additionally, the general trend in the coating industry can be described by the replacement of thick multilayer coatings with thinner, less layer containing and highly efficient coatings systems.
The good adhesion and corrosion protection properties of epoxy coatings predestine these coating materials to be excellent primers but, because of their lack in UV- and weathering stability, polymers on the base of acrylics or polyurethanes must be used as top or clear coats. In principle, these coating layers can be applied as liquid or powder coating systems. Besides the fact that powder coatings can be a cost effective alternative in most of the coating applications, it cannot be ignored that they can be applied only at a higher layer thickness and very often, if compared to their liquid coating counterparts, additionally a lower coating performance is observed.
To bridge the gap between the coating market demands for cost efficient, single layer application coating systems and material or processing cost savings on the one hand and the need for highly durable, protective coatings on the other hand, will be one of the challenges in the future coating industry. It is therefore obvious, that an urgent need for new smart powder coating systems, which combine in a cost efficient way high durability and a good performance, with effective protective properties, exists.
In the presented study, an innovative concept for smart coatings is proposed, which is based on stratification effects in epoxy and polyurethane components containing powder coating formulations.

Keywords: stratifying; smart powder coating;

S.31-3 Lumiflon feve based formulation for hyper durable powder coatings

Floris Kooistra
Technical Service, AGC CHEMICALS EUROPE, Amsterdam, the Netherlands

Fluoro Ethylene Vinyl Ether (FEVE) resins were commercialized by AGC in 1982 under the brandname Lumiflon as the first soluble fluorinated resins for coating formulations. Nowadays a wide portfolio of Lumiflon resins is available in different forms, such as waterborne emulsions and dispersions, classical solvent borne and powder form.
FEVE resins are renowned for their extremely high durability due to the high C-F bond energy. Therefore, they do not degrade under the influence of UV radiation from the sun. Coatings based on FEVE resins show outdoor durability performance of over 30 years.
FEVE based powder coatings are widely used for powder coatings which need to fulfil the toughest architectural specifcations like AMAA 2605 and Qualicoat Class 3. These specifications require 10y Florida exposure performance showing gloss retention >50% and colour change < 5 ΔE. Meeting these requirements today is only possible with fluorinated resin technology.
This paper discusses the basic chemistry behind the extreme durability of Lumiflon based coating systems. Furthermore, accelerated weathering test methods are compared and the paper shows the excellent correlation between HE-XWM and outdoor exposure in Florida and Okinawa. HE-XWM testing focusses on degradation caused by the photocatalytic effect of TiO_2 and is able to predict 10y outdoor weathering after only 200 hours of testing. Using this fast test allows for quick formulation development.
The paper shows how to optimize a hyper durable powder coating formulation by choosing the right raw materials. Blending the FEVE resin with hyper durable polyesters can create a self-stratifying coating forming a thin clear coat of fluorinated resin at the surface. This greatly enhances the durability of the system which is supported by accelerated weathering testing and actual 10y Florida outdoor exposure data. Also, images from cross cut microscopy will support this finding. Next to the resin selection the paper shows data on choosing the right TiO_2 grade, NCO hardener and UV/HALS additives.
In conclusion, Lumiflon FEVE resins can be easily used for formulating hyper durable powder coatings meeting Qualicoat class 3 or AMAA 2605 standards. A fast accelerated weathering test focussing on the photocatlyic degradation effect of TiO_2 allows formulators to quickly optimize their formulation. Thereby greatly reducing development time of hyper durable powder coatings.

Keywords: durability, FEVE, HE-XWM, lumiflon

S.32-1 Film formation of latex emulsions followed by white light interferometry.

Alberto Olivares[1], Julian Galvan-Miyoshi[1], Augusto García-Valenzuela[2], Fernando Zaldo[3]
[1]CIP- Applied Physics, PPG, Mexico
[2]CCADET, UNAM, 3CIP, PPG , Mexico

Reflectance spectra at normal incidence were recorded from different latex applications during their film formation processes at ambient temperature. Besides oscillations in the reflected spectra due to two-wave interference phenomenon in an effective film, a clear asymmetric Fano-like resonance appeared during long periods of time in some of the latex films studied. We develop an optical model to explain all the observed interference phenomena and elucidate in simple terms the underlying mechanism of the observed Fano resonances. The model reproduces qualitatively very well all the features in the experimental reflectance spectra. We find that the Fano resonances arise due to the interference between the surface reflection and a Bragg reflection from self-assembled photonic crystals within the film. This technique has been applied to the study of different latex particles with different softness and the interaction potential.

References:

1. *Olivares, J. A., et al. "Fano-type resonances in the reflectance spectra from dense colloidal films." Optics Communications 356 (2015): 175-181*
2. *J.A. Olivares, J.M. Galván-Miyoshi, F.L.S. Cuppo, F. Zaldo, A. García-Valenzuela, "Drying dynamics of polymers films by an interferometric technique", Prog.Org. Coat. 77 (2014) 1531–1538*

Keywords: film formation, latex drying, light interferometry

S.32-2 Waterborne fluoropolymers: a breakthrough alternative to conventional matting agents

Silfredo Javier Bohorquez[1], Ana Belen Lopez[2], Marcel Meeuwisse[1], Dirk Mestach[1], Jose C de la Cal[2], Jose M Asua[2]
[1]Allnex, Bergen op Zoom, the Netherlands
[2]POLYMAT, San Sebastian-Donostia, Spain

Nowadays, there is a continuous demand for the use of functional coatings that have an intrinsically matt appearance. For instance, in the industrial wood industry for some applications it is desired to bring out the natural structure of the wood (a coated piece that looks naturally uncoated). Different commercial agents that are based on treated silica, polyethylene waxes and/or micronized organic polymers are available, however; the compatibility of the matting agents and their incorporation into the paint formulation with the waterborne binder is usually complex and tedious. Moreover, the matting agents in most of the cases do not bring extra properties to the films. In that sense it would be ideal to have a matting agent that has the similar nature of a waterborne acrylic binder, in terms of compatibility, film formation and performance.

This work describes a study on the incorporation of waterborne fluoropolymer dispersions into paint formulations to act as a film forming binder for matting and chemical resistance performance purposes. The waterborne polymer were synthesized using processes described elsewhere[1,2]. Different strategies were used to incorporate the waterborne fluoropolymers into the paint formulation, which also included the partial substitution of commercial matting agents. AFM and SEM techniques were employed to evaluate the stratification of the fluoropolymer particles and the surface of the films. The pigmented paints were applied onto different substrates and the matting efficiency and the chemical resistance performance evaluated. The results clearly showed that these waterborne fluoropolymers can be used as an alternative replacement for commercial matting agents, while adding value to the coating by offering enhanced hydrophobicity as compared to regular acrylic binders.

References

1. Asua, J. M. Miniemulsion Polymerization. Prog. Polym. Sci. 27 (2002) 1283–1346.
2. López, A. B.; de la Cal, J. C.; Asua, J. M. Controlling Film Topography to Form Highly Hydrophobic Waterborne Coatings. Soft Matter 12 (2016) 7005–7011.

Keywords: fluoropolymers, self-matting, waterborne

S.32-3 Sailing the uncharted waters of 2k polyurethane coatings

Frederic Lucas, Sebastian Roller, Ulrich Tromsdorf, Manfred Dargatz, Rabie Al-Hellani, Quirin Kainz
BASF, Ludwigshafen, Germany

2K waterborne polyurethane coatings present an attractive solution in order to achieve high performance while working with a sustainable system. Indeed, they offer a unique possibility to reach the outstanding properties of polyurethanes systems coupled with the sustainability inherent when working with water. The development of raw materials compatible with aqueous formulations led to the development of water-emulsifiable polyisocyanates. Their synthetic approach is based on the modification of a standard aliphatic polyisocyanate crosslinker with a so-called "hydrophile".

As part of our aspiration to unravel the mysteries of 2K waterborne polyurethane coatings, we spent time and effort to understand the influence of the hardener structure on the final lacquers properties. In the sake of gaining a better insight, the elucidation of the phenomenon occurring in the aqueous formulation also had to be tackled. While offering a good insight into the film formation of in 2K waterborne polyurethane systems these results also raised some unexpected questions.

In a first part, the challenges and the key features related to the hydrophilization of the polyisocyanate were investigated. More precisely, the two-main types of hydrophiles (steric hydrophile such as OH functional poly (ethylene oxide) and ionic hydrophiles) were compared. As could be expected, both approaches offer advantages and drawbacks and no technical solution should be seen as the holy grail. Thus, the nature of hydrophile should be selected wisely depending on the targeted application.

In a second part, the behavior of the hardener when formulated in the aqueous binder (generally a polyacrylate or a polyurethane dispersions) was studied. This work was rendered difficult by the fact that the isocyanate function can undergo a wide variety of reactions which compete with each other. Additionally, the system is heterogenous and factors such as accessibility and diffusion had to be taken into consideration. As already published 1,2, density gradient ultracentrifugation enabled the understanding of the main phenomena occurring upon storage. It was observed that the fate of the crosslinker strongly correlates with its partitioning and its reactivity which both tremendously influence the pot-life of the formulation.

Finally, the focus was laid on the impact on the polyisocyanate on the performance (and more particularly the gloss) of the corresponding coatings. Coupling property measurements with analytical investigations (AFM, FTIR or NMR), it was observed that the efficiency of the emulsification, the solvent used and the compatibly of the crosslinker in the system are the main leverages to achieve high gloss.

References:

1. *F. Lucas et al, Farbe und Lack, 2017, Volume 123, Issue 8, Pages 20-24*
2. *F. Lucas et al, European Coatings Journal, 2017, Issue 11*

Keywords: film formation, film properties, polyurethane coating, waterborne coating

S.33-1 On the way to biocide-free anti-fouling and foul-release coatings

Katrin Roland, Sascha Herrwerth
Evonik Resource Efficiency GmbH, Essen, Germany

Shipping accounts for 90 % of the global goods transportation. The economic viability of the global merchant fleet depends heavily on the shell of the vessels. Biofouling, caused by plant and animal growth on the hull under the waterline, is a particular concern. Proteins, carbohydrates, and other micronutrients from the ocean attach themselves to the hull only seconds after its first contact with seawater. As nutrients, they attract marine organisms that settle on the surface, which is converted into a complex habitat in which several hundred species interact with each other. This increases the surface roughness and thus the friction during the vessel's movement. Consequently, to maintain the same speed of travel, the ship needs more energy and fuel, with an associated rise in CO_2 emissions. Hence, the demand is high for coatings solutions that protect ships' hulls from biofouling.

Today, the most effective way to reduce biofouling are biocidal antifouling paints. These are based on a leaching process, which continuously releases poison to the seawater, killing all organisms in the surrounding of the ship, before they can settle on the surface. However, for environmental reasons and increased regulatory pressure, there is a high demand for biocide-free antifouling or foul release coatings.

Starting out from an established product class, an amphiphilic silicone-hybrid resin was developed, which can be used as a binder for foul-release coatings. A hydrophobic silicone-epoxy resin, with its strong anti-adhesive properties, was combined with a hydrophilic polymer. On contact with water, the hydrophilic moieties will develop a hydrate layer around the polymer, and therewith create a film of water on top of the surface. This layer will mask the ship's hull from surrounding organisms and they will refrain from settling on the solid surface. Furthermore, the hydrophobic silicone backbone adds an easy-to-clean effect to the coating.

The foul release properties of this binder were tested in newly developed laboratory tests. In one method, a mesocosm was created in a special aquarium. Over a period of six weeks, the coated panels were exposed to conditions comparable to an algal bloom under static conditions. This was followed by one week of rotation, simulating a moving vessel and releasing loosely attached organisms. In addition, a microcosm test was carried out, employing co-cultures of diatoms and bacteria. The results from both tests were in accordance with the results from seawater exposures, indicating that the amphiphilic silicone-hybrid polymer can serve as a binder or co-binder for foul-release coatings.

As an intermediate step on the way to biocide-free foul-release coatings, a structure-modified Fumed Silica can assist to reduce the Cu_2O content of biocidal antifouling coatings dramatically. It was shown, that the silica particle forms a strong complex with the cuprous oxide, modifying its surface. Hence, the leaching rate of the biocide is reduced, avoiding overdosing of the poison into the marine environment. Formulations with very low Cu_2O content are currently exposed in the North Sea. After more than two years, the performance is still comparable with higher biocide loadings.

Keywords: anti-fouling coatings

S.33-2 Nanobiocoatings - invisible particles with the visible effects

Dušan Fašung
Coating services, Top Coatings, Suchohrad, Slovakia

Nanotechnology is changing the properties of materials at the level of atoms and molecules.
It is also widely used in long-term protection of materials as: stone, wood, glass, metal, plastic,textiles in interior and exterior paving, tiles, facades, industry coating, solar PV pannels... Engine additives
Architects often use in modern architecture modern materials and exception are not even facades of buildings and structures.
Unfortunately, most of the material is indeed attractively designed but functionally are not capable of withstanding external environmental influences. UV radiation, acid rains, dust, smog, salt, low or high temperatures. These are all negative factors affecting the durability and maintenance of facades.
The existing protection based on silicones, waxes, acrylics and various other coatings showed up in today's urban time insufficient. Often change color, do not have sufficient resistance to UV radiation and in the results comes to its degradation, which brings lowering aesthetics and functional mission. Leads to overheating of the building skin, it leds to the increasing energy demand for air conditioning and the general deterioration of the internal and external environment. Overheating is a common phenomenon that is on the agenda urbanists, ecologist and planners.
The latest research in nanobiotechnology has enabled us to develop coating and we are already industrially manufacture and used in practice. Change the properties of materials at the level of atoms and molecules, thus eliminating external influences, while the materials retain their shape, color and other characteristics. This reduces dilatation of the material, which is of great importance especially in ventilated, but also glued facades. This reduces the risk of damage due to frost insulation, even salt and water increases the volume by 8%.
Great benefit for developers is reduced maintenance costs because all materials that are treated with nanobiocoatings are UV-resistant, water-resistant, antistatic, self-cleaning and any rain just flushed the building.
We are working with particles of 3 nanometers, which create a protective grid of 3 x 3 nanometers. Since all viruses, bacteria and spores of moss and algae are larger than the specified protection grid, they cannot in so treated surface to reproduce and live. On the facades will not form colonies of moss and algae. The protection is not toxic, but essentially mechanical.We are environmentally friendly.

References:

1. *Bory Mall, Bratislava, Slovakia*
2. *Hotel Kempinski, High Tatras, Slovakia*
3. *Hyatt Capital Gate Hotel, Abu Dhabi, UAE*
4. *Burj Rafal Kempinski Hotel, Riyadh, Saudi Arabia*
5. *Hotel Sheraton, Bratislava, Slovakia*

Keywords: Nanobiocoating, Longterm duration, UV-resistant, antistatic, selfcleaning, anticorrosion,

S.33-3 The development of a UV-LED fouling prevention system for ship hulls

C Price[1], Kevin Reynolds[1], N Hijnen[1], C Cairns[1], M.F Van der Zalm [2], W Widjaja[2], F Loh[2], R Hietbrink[3], M Jongerius[3], M Wijnen[3]
[1]AkzoNobel Marine and Protective Coatings, Gateshead, United Kingdom
[2]AkzoNobel Marine and Protective Coatings, Tuas, Singapore
[3]Philips IP&S, Eindhoven, the Netherlands

AkzoNobel and Philips are jointly developing a pioneering fouling prevention system based on UV-LED technology; by combining AkzoNobel's coating formulation, surface protection and adhesion know-how with Philips' capability in UV-LED lighting and electronics, the two companies are aiming to develop an economically viable solution for the prevention of fouling settlement on ship hulls and underwater surfaces.The approach reported herin is based on the use of ultraviolet (UV) light to control fouling and prevent biological roughness which increases the hydrodynamic drag, fuel consumption and emissions of marine vessels. UV light has long been known to kill bacteria and is a common method for water sterilization in the medical and food industries. This new technology integrates low intesity UV LEDs in a protective film scheme which is designed to allow the UV light to be emitted from the surface, thus providing the total prevention of biofouling accumulation on the protected area. The system which is completely free of biocides and uses low intensity LEDs only exposes fouling organisms in very close proximity to the surface to a harmful dose of UV light.This new innovation provides complete fouling prevention to ship and boat hulls; this total control of biofouling delivers substantial economic and environmental benefits and represents a step change improvement in fouling control performance vs. current state of the art systems. A 11,000 TEU container vessel using this approach is predicted to save over $7m in fuel costs over a 5 year dry dock period compared to a high performance Self-Polishing Copolymer (SPC) antifouling. This corresponds to a fuel consumption saving of 18,773 metric ton of bunker fuel and 59,135 metric ton of associated CO_2 emissions.The technology is based upon a flexible, waterproof and optically transparent silicone laminate tile which can be applied to the outer hull of a vessel over existing primer schemes. In-lab bioassay testing has been performed along with field testing in a range of environments including those with a severe fouling challenge such as Singapore. Testing has progressed to boat and ship patch trials with a program designed so that fouling prevention will be demonstrated across a range of environmental and operational conditions.The first results demonstrate proof of concept (total fouling prevention) whereby test panels have been shown to remain completely free of any fouling growth. This presentation details both the laboratory and in-field fouling prevention testing of the latest prototype devices and outlines the forthcoming stages of product development including materials selection, scale-up, test patching and vessel trials. We will also provide the most recent insights and advances on the electronics and optics, the UV-LED technology, the powering approach, and the mechanical design.

S.34-1 Solutions to address future EU regulatory restrictions on MIT

Emmanuelle Yvon
Ressearch, Dow Europe GmbH, Horgen, Switzerland

In the coming years, significant changes to European legislation will impact European paints and coatings manufacturers.

The proposal of the ECHA risk assessment committee (RAC) to lower the specific concentration limit (SCL) for MIT (2-methyl-2H-isothiazol-3-one) from 1000 ppm down to 15 ppm represents a tremendous challenge for the paints & coatings Industry that is expected to become reality by 2019-2020 at the latest.

Finished products with concentrations higher than 15 ppm of MIT will have to be labelled with the CLP hazard statement H317 "May cause an allergic skin reaction". As a consequence of this new CLP implementation, paints containing ≥ 15 ppm MIT will no longer be compliant with European ecolabel. Moreover, product containing more than 15 ppm of MIT are expected to get banned for consumer uses toward 2022-23.

Dow is in position to help and support the paints & coatings industry to overcome this challenge with its new biocidal active substance MBIT. MBIT is an efficient and sustainable biocide that offers a suitable technical alternate for MIT replacement in preservation systems without triggering H317 labelling at dose levels.

This paper will provide an overview of the latest regulatory dynamics and challenges to the industry, present a comparative technical assessment of MBIT vs. MIT and present how high through put testing technology helps to develop, optimize and validate tomorrow's preservation systems.

References:

1. *https://echa.europa.eu/registry-of-submitted-harmonised-classification-and-labelling-intentions/-/substance-rev/15343/term.*
2. *https://www.echa.europa.eu/web/guest/harmonised-classification-and-labelling-previous-consultations/-/substance-rev/3394/del/50/col/synonymDynamicField_622/type/desc/pre/2/view*
3. *Askew P D (2006) Coatings and Coatings Materials in Springer Handbook of Materials Measure-ment Methods Eds Czichos H, Saito T, Smith L, Springer Science and Business Media*
4. *Lunenberg-Duindam J, Lindner W (2000). In-Can Preservation. European Coating Journal. 03, 66 -73.*
5. *Breakthrough innovation for in-can preservation, PPCJ August 2011, Anders Carlsen and Gerhard Tiedtke, Dow Microbial Control, Switzerland.*
6. *New Biocidal Active, PCI April 2014, Beth Ann Browne, Ph.D., Dow Microbial Control, US.*

Keywords: biocide, MIT, preservative, regulation

S.34-2 Advanced solutions for transparent iron oxide pigment dispersions in waterborne concentrates

Loulou Rozek
Technical Service, Borchers, Westlake, USA

For many years, paint and colorant manufacturers faced difficulties dispersing transparent iron oxide pigments in aqueous systems. Long grind times, high grind viscosity, poor stability and low film transparency are typical challenges that kept these types of pigments from expanding into the rapidly growing waterborne wood and automotive coatings. This paper introduces a brand-new dispersant, designed and engineered to produced strong and fast interactions with the surface of these pigments allowing for a fast drop in viscosity during the dispersion process as well as film transparency that rivals those of solvent-borne coatings.
In this paper, four leading grades of red and yellow transparent iron oxide pigments are evaluated using the novel dispersant and compared to the industry benchmark. A correlation between the particle size distribution of the grinds and its impact on transparency as well as viscosity and stability of the dispersion will be discussed. Additionally, a method for quantifying transparency is demonstrated using 1k acrylic and 2k acrylic binder systems.
Producing and applying waterborne transparent iron oxide pigment dispersions is now possible and easier than ever before.

S.34-3 Reactive coalescing agents (RCAS) towards low VOC coatings

Satyasankar Jana, Ranganathan Krishnan, Jagjit Kaur, Phang Jia min, Marco Klähn, Balamurugan Ramalingam and Alexander van Herk
Polymer Engineering and Characterization, Institute of Chemical and Engineering Sciences, Singapore, Singapore

One of the successful industrial approaches of making surface coatings with low environmental impact[1] is adopting water based technology such as using waterborne formulations instead of solvent based ones. Waterborne coatings are widely used as architectural and industrial coatings, paper coatings etc. In general, emissions from waterborne coatings are significantly less in compare to the solvent based coatings due to the presence of water as dispersing media. However, present waterborne coatings still use substantial amount of small organic compounds (SOC) or volatile organic compounds (VOC) as functional additive for different purpose. Coalescing agent (CA) is the main SOCs/VOCs used in present waterborne coating formulations. The role of a CA is to enhance film formation by decreasing glass transition temperature (Tg) or minimum film formation temperature (MFFT) of the film forming polymer and increasing polymer chain diffusion during coating drying process. However, these CAs get evaporated into atmosphere once the coating is dried, which causes environmental pollution and leads to long term health hazard. One of the novel strategies to overcome this issue is to use of reactive coalescing agents (RCAs).[2] These RCAs act as normal CAs until the coating is dried. However, once the drying process is completed they react themselves or react with other components of the coating and become an integral part of the coating for ever.
In this presentation, at first, an overview and recent status on RCA based technologies will be presented. Molecular Dynamics (MD) simulations, supported by experimental results, to establish the effectiveness of monofunctional RCAs on Tg change of the film forming polymer will be reported. We will also report synthesis, characterization and reactivity evaluation of a few i) (meth)acrylate based RCAs and ii) novel hydroxyethyl sulfone based RCAs that shows water-release-triggered[3] reactions. Coalescing performance of these RCAs were investigated by measuring MFFT of RCA blend latices and reactivity of these RCAs were evaluated by using relevant trigger in presence of film forming polymer. Finally, a comparative study on emission profile of these RCAs with their unreactive counterparts using head space gas chromatography (HS-GC) will be presented.

Acknowledgement
This project is funded by Agency for Science, Technology and Research (A*STAR), Singapore under NoSOC programme.

References:

1. *https://www.coatingsworld.com/issues/2016-09-01/view_features/recent-trends-in-architectural -coatings/*
2. *Rohm & Haas Co., EP1106593 B1, 1998;*
3. *Rohm & Haas Co., US6509294 B1, 2003;*
4. *Archer Daniels Midland Co., WO2007094922 A2, 2007;*
5. *Rohm & Haas Co., EP1447432A1, 2004;*
6. *J. V. Barbosa, E. Veludo, J. Moniz, F. D. Magalhaes, and M. M. S. M. Bastos, Eur. J. Lipid Sci. Technol., 2012, 114, 1175;*
7. *Valtion Teknillinen Tutkimuskeskus, WO 2005090501 A1, 2005;*
8. *M. Lahtinen, E. Glad, S. Koskimies, F. Sundholm, K. Rissanen. J. Appl. Polym. Sci., 2003, 87, 610-615.*
9. *D. J. Berrisford, P. A. Lovell, N. R. Suliman, A. Whiting, Chem. Comm., 2005, 5904-5906.*

Keywords: coalescing agent, low VOC coatings, minimum film formation temperature, gas chromatography

S.35-1 Observing the fabricating of nanostructured conductive ink coatings in situ using x-rays

Stephan Roth[1, 2], Björn Fricke[2], Calvin Brett[2, 3], Daniel Söderberg[3], Wiebke Ohm[2], Matthias Schwartzkopf[2], Tim Laarmann[2, 4]
[1]Fibre and Polymer Technology, KTH Royal Institute of Technology, Stockholm, Sweden
[2]Deutsches Elektronen-Synchrotron, Hamburg, Germany
[3]Department of Mechanics, KTH Royal Institute of Technology, Stockholm, Sweden
[4]CUI, Hamburg, Germany

The installation of nanostructured and conductive coatings based on metal nanoparticles inks is crucial for the fabrication of coatings with advanced properties. Inks containing colloids or precursors can be deposited by a multitude of methods, such as ink-jet printing, slot-die coating and spray coating. Among them, spray deposition is a scalable, rapid, and roll-to-roll compatible coating method (1). It offers unique opportunities for tailoring the morphology of the coated layers by tuning the spray parameters, such as droplet size, evaporation rate and temperature (2). Here, we combine spray coating with metal-organic decomposition inks (MOD) (3). MODs offer a facile route for preparing functional nanoparticle and conductive coatings based on precursor and solvent-based methods. Aim of the present study was to investigate the spray coating process itself and the subsequent post-treatment of the deposited ink layer on a hydrophilic silicon oxide layer.

To do so, we studied the spray deposition of a model ink consisting of a silver nitrate precursor in ethanol solvent due to its solubility and the low toxicity and analyze the subsequent transformation into metallic silver layers via thermal annealing in situ and in real-time. We employed a combination of time-resolved surface-sensitive X-ray scattering, from the atomic to the nanoscale, and optical spectroscopy (UV-Vis) to observe the spray-coating process at the substrate itself as well as the subsequent layer formation and transformation due to post-deposition thermal treatment. In addition, the resulting layers were subsequently imaged by optical microscopy and atomic force microscopy. This unique combination allowed us to identify morphological and chemical changes during the spraying and heating process, in particular the crucial influence of the spray parameters on the layer morphology was elucidated. The average volume of the deposited silver nitrate nanoparticles is increased at nearly constant surface density. Thus the silver nitrate covers the silicon oxide surface.

During the heat-induced transformation, the silver nitrate precursor nanoparticles undergo a two-fold transformation. On the nanoscale, deliquescence is observed, resulting in a larger percentage of surface coverage in comparison to the as-deposited films. Additionally, the temperature dependent conversion on the atomic scale from silver nitrate to metallic silver is detected, already at a lower temperature of approximately 140°C, being much lower than expected and being clearly influenced by the layer thickness in the submicron range. The transformation is corroborated by the observation of a plasmon resonance, a characteristic optical absorption band in metallic nanoparticles. This is indispensable for optimizing the coating by tuning the deposition and annealing parameters for future use as ink-based sensors, catalysts and conductive coatings.

References:

1. *Roth, S. V. J. Phys. Condens. Matter 2016, 28 (403003), 37.*
2. *Amirav, L., Lifshitz, E. ChemPhysChem 2015, 16 (353), 7.*
3. *Shin, D.-Y., Jung, M., Chun, S. J.Mater.Chem. 2012, 22 (11755), 10.*

Keywords: ink, nanostructure, spray coating, thin film

S.35-2 Tailoring of optical properties of polymer thin films via metal sputter deposition

Matthias Schwartzkopf [1], Calvin J. Brett[1, 2], Oleksandr Polonskyi [3], Alexander Hinz[3], Thomas Strunskus[3], Franziska C. Löhrer[4], Volker Körstgens [4], Franz Faupel[3], Peter Müller-Buschbaum [4], Stephan V. Roth[1, 2]
[1]DESY, Hamburg, Germany
[2]KTH, Stockholm, Sweden
[3]CAU, Kiel, Germany
[4]TUM, München, Germany

Metal nanoparticles on polymer thin films promise various potential applications as functional coatings in solar cells, biosensors, as reflective or antireflective coatings, and in heterogeneous catalysis [1]. Fabricating such nanocluster films with tailored morphology, their characterization and manipulation at the nanoscale are the essential prerequisites and tools for implementing devices in all of these fields. Typically chemical routes allow for a large variety of nanoparticles with different size, shape and composition; here, we exploit physical vapor deposition as a complementary approach for physical bottom-up fabrication. Sputter deposition plays a very important role in industrial and nanostructural processing to reproducibly tailor the average nanoparticle size and distances with high surface coverage demanded for efficient device performances. In order to tune the size-dependent optoelectronic properties, it is mandatory to monitor how the growth kinetic affects the metal film morphology and how it correlates to the optical properties during sputter deposition. To obtain full control over the nanostructural evolution at the metal-polymer interface and its impact on optoelectronic properties, we employed a combination of in situ time-resolved surface-sensitive X-ray scattering with in situ optical Spectroscopy during sputter deposition of gold (Au) on thin polystyrene films (PS) [2]. We monitored the evolution of the metallic layer morphology according to changes in the key scattering features by geometrical modeling [3] and correlate the nanostructural development to optical propeties. This enables us to identify the different growth regimes including their specific thresholds and permits better understanding of the growth kinetics of gold clusters and their self-organization into complex nanostructures on polymer thin films.
During sputter deposition, a change in optical reflectivity of the pristine grey-blue polystyrene film occurred ranging from dark blue color due to the presence of isolated nanoclusters at the interface to bright red color from larger Au aggregates [4]. Furthermore, a surface diffusion coefficient according to the kinetic freezing model and interfacial energy of Au on PS at room temperature were calculated based on a real-time experiment. We report in a recent study the role of sputter deposition rate in tailoring metal layer morphologies on polymer thin films [5]. The deposition rate affects primarily the nucleation process and the adsorption-mediated growth, whereas rather small effects on diffusion-mediated growth processes are observed. Only at higher rates, initial particle densities are higher due to an increasing influence of random nucleation and an earlier onset of thin film percolation occurs. Thus, our findings are of great interest for applications in organic photovoltaics and organic electronics, which benefit from tailored metal-polymer interfaces.

References:

1. *Faupel F, Zaporojtchenko V, Strunskus T, Elbahri M, Adv. Eng. Mater. 12, 1177 (2010).*
2. *Schwartzkopf M and Roth SV; Nanomaterials, 6, 239 (2016).*
3. *Schwartzkopf M, Buffet A, Körstgens V, Schlage K, Perlich J, Rothkirch A, Benecke G, Metwalli E, Heidmann B, Rawolle M, Herzog G, Müller-Buschbaum P, Röhlsberger R, Gehrke R, Stribeck N und Roth SV; Nanoscale, 5, 5053-5062 (2013).*
4. *Schwartzkopf M, Santoro G, Brett CJ, Rothkirch A, Polonskyi O, Hinz A, Metwalli E, Yao Y, Strunskus T, Faupel F, Müller-Buschbaum P und Roth SV; ACS Appl. Mater. Interfaces, 7, 13547–13556 (2015).*
5. *Schwartzkopf M, Hinz A, Polonskyi O, Strunskus T, Löhrer FC, Körstgens V, Müller-Buschbaum P, Faupel F und Roth SV; ACS Appl. Mater. Interfaces, 9, 5629–5637 (2017).*

Keywords: metal - polymer interface, polymer thin films, sputter deposition, X-ray characterization

S.35-3 Fabrication of an epoxy system with excellent anti-corrosion properties through impregnation of cerium oxide decorated graphene oxide (go-CeO_2) nanocomposite

Bahram Ramezanzadeh, Zahra Sanaei, Mohammad Ramezanzadeh
Institute for Color Science and Technology, Tehran, Iran

A high performance anti-corrosion system based on epoxy coating was fabricated through inclusion of GO-CeO_2 nanosheets into the epoxy coating. First, the CeO_2 nanoparticles were synthesized on the GO-PAni surface through an in-situ process. The chemistry and morphology of the GO-CeO_2 nanosheets were characterized by Fourier transform infrared spectroscopy (FT-IR), X-ray photoelectron spectroscopy (XPS), X-ray diffraction analysis (XRD), thermal gravimetric analysis (TGA), UV-Vis spectroscopy, and field-emission scanning electron microscopy (FE-SEM). The GO-CeO_2 nanosheets were incorporated into the epoxy matrix, applied on mild steel, immersed in chloride solution (3.5 wt.%), and characterized by salt spray and electrochemical impedance spectroscopy (EIS). Results revealed that the corrosion protection properties of epoxy coating significantly enhanced by addition of GO-CeO_2 nanosheets. The good barrier performance of GO sheets due to their high specific surface area, and the cathodic sites protection by Ce4+ cations are responsible for the epoxy coating anti-corrosion properties enhancement in the presence of GO-CeO_2 nanocomposites.

Keywords: anti-corrosion, CeO_2 nanoparticles, epoxy, graphene oxide nanosheets

S.36-1 Functional latex dispersions: from superhydrophobic to antimicrobial coatings

W. Marshall Ming
Department of Chemistry and Biochemistry, Georgia Southern University, Statesboro, USA

Polymer latexes have been widely applied as architectural coatings. Incorporation of functional groups into polymer latexes would endow various functionalities to water-borne coating. As the first example, we have designed and prepared silicone-containing acrylic latexes from emulsion polymerization, which upon film formation led to coatings with surface wettability similar to a silicone surface and bulk properties close to acrylic polymer. By judiciously combining silicone-containing latexes with micro/nano-sized inorganic particles, we obtained robust water-borne superhydrophobic coatings with high advancing water contact angles and low roll-off angles. In the second example, we synthesized polymer latexes with covalently bonded, hydrophobic quaternary ammonium compounds (QACs), and upon film formation we obtained non-leaching antimicrobial coating as demonstrated by up to 5-log reduction against both Gram-positive S. epidermidis and Gram-negative E. coli. The strong antimicrobial activity is due to the significant surface enrichment of QACs, and the antimicrobial action is based entirely on contact killing, without leaching of bactericidal species, as revealed by the zone-of-inhibition test. The water-borne antimicrobial coating may find wide applications in high touch, high risk environments.

S36-2 Designing coatings to have low adhesion surfaces through self-stratification

Dean Webster
Coatings and Polymeric Materials, North Dakota State University, Fargo, USA

Designing coatings to have important functions is an ever-increasing area of coatings science. Many applications of coatings require systems that have surfaces tailored to deter adhesion to their surfaces. These applications include low dirt pick-up coatings, anti-graffiti coatings, biomedical coatings, anti-icing coatings, and non-fouling and fouling-release marine coatings. Using the concept of self-stratification, it is possible to decouple the surface and bulk properties of coatings so that the bulk and surface properties can be tuned individually to meet the material requirements. The approach of self-stratification has been applied to the design of mechanically robust marine fouling-release coatings. By combining functional poly(dimethyl siloxane) (PDMS) with a polyurethane system, coatings having the bulk properties of polyurethanes (good adhesion, toughness) as well as low surface energy can be designed that have good fouling-release performance for a wide variety of marine organisms. However, there are some marine organisms that are able to adhere well to a low surface energy surface such as PDMS. To overcome this and provide a coating having broad-spectrum low adhesion and amphiphilic approach has been investigated. First, a prepolymer was synthesized by the reaction of a polyisocyanate with a monofunctional PDMS and monofunctional poly(ethylene glycol) (mPEG). The prepolymer was mixed with an acrylic polyol, polyisocyanate and coatings prepared. The coatings were characterized using contact angle measurements, AFM, and XPS. The coatings were also subjected to a series of laboratory assays using a variety of marine organisms including diatoms, green algae, barnacles, and mussels. Clear trends could be observed in the barnacle adhesion as a function of PDMS and PEG chain length and coatings having good fouling-release performance for a variety of marine organisms were identified. The coatings were also subjected to ice adhesion tests and a very interesting and rather direct correlation was observed between the adhesion strength of ice and barnacles, indicating that the same design rules for designing fouling-release coatings could be applied to the design of ice-release coatings.

Keywords: anti-adhesion, fouling-release, marine paint

S.36-3 Rheological study of the effect of additives on the drying of waterborne dispersion paints

Jesus Pitarch[1], Joerg Rueger[1], Christian Sproncken[1,2], Marianne Gravendeel[1,3], Klaus Oldörp[4], Cornelia Küchenmeister-Lehrheuer[4]
[1]Clariant, Frankfurt am Main, Germany
[2]TU Eindhoven, Eindhoven, the Netheralnds
[3]TU Delft, Delft, the Netherlands
[4]Thermo Fisher Scientific, Karlsruhe, Germany

A simple rheological method based on a novel measuring system is introduced which allows the investigation of the drying behaviour of waterborne dispersion paints. The new system is based on twelve pins equidistanly fixed on a ring geometry, which is attached to a highly precise air-bearing rheometer. The pins penetrate in a wet paint film of defined thickness and slowly rotate through it. The detected change of torque over time is directly related to a change of film viscosity and allows therefore to detect the formation of a high viscous close-packed arragement of particles as a characteristic transition during drying, which is assigned to the end of the open time.
It will be shown how this method has been used to characterize the drying behaviour of polymer dispersions and paints and to differentiate the performance of additives regarding their impact on film drying. The new test system will help formulators to select in an easy and objective way the most favourable ingredients to optimize the drying behaviour of aqueous paints.

Keywords: additives, open time, rheology, waterborne coating

PLENARY LECTURE 3

PL.03 Current and future technologies for the coating industry

Jürgen Omeis[1], Guillaume Jaunky[2]
[1]Innovation, ALTANA AG, [2]R&D, BYK-Chemie GmbH, Wesel, Germany

Abstract content: Current and Future Technologies for the Coating Industry
Jürgen Omeis, ALTANA AG
Guillaume Jaunky, BYK-Chemie GmbH
The world of coatings, vanishes and inks is changing continuously. From a general view there are several trends driving this change. Besides the strong growth of the economy in the eastern part of the world especially in China increasing global legislation and request for sustainability have a big impact on the coating industry and technology. Of course these driving forces are always challenged by commercial constraints. How does the coating industry cope with these challenges and where are the chances for the future? Obviously there is no single answer but a lot of efforts to push technology in order to find better solutions for today's requirements and look for new approaches fit for the future. General performance requirements for coatings are as diverse and "colorful" as the industry itself. Currently surface protection, perfect image and life time are most common followed by specific demands depending on the industry, ranging from decorative paints to automotive and electronic applications. Sustainability is a key-word which is often related to water based coatings and renewable raw materials and building blocks. Beyond this existing trend and technology there are other options for the future providing new features to coatings like self-cleaning or selfrepair properties or new concepts for antifouling and dirt pick-up. There are recent studies to adapt bionic principles with new surface materials to generate new effects which might be suitable for coatings if robustness is improved. Aligned with this new technology pathway are triggered systems which can switch actively properties by an external stimulus. Legislation additionally is a challenge and chance for the future. Worldwide activities ranging from chemical inventories like REACH, MITI, TSCA to stringent regulations for food contact and others put severe restrictions on the use of hazardous chemicals. Prominent examples are the Bisphenol A ban for can coatings and the substitution of C8-perfluoro-compounds. Moreover there are growing constraints for heavy metal catalyst, biocides and toxic co-solvents or monomers. Alternative technologies need to be developed or improved for the entire range of components. Especially in the additive and effect pigment world there are already projects targeting the next level. Last but not least performance features themselves or in combination with application technologies are an evergreen. This comprises a whole range of technology topics like new materials and processes for barrier applications, digital 2 and 3 D coatings and applications, new security features, haptics and other effects as well as printable electronics. The combination of application technology, electronic process control and new materials will give further potential for disruptive approaches in coatings, printing and packaging industry. Of course, decorative topics like color and image will always play a role today and in the future for the end customer. The content of the paper will cover the major trends in general and provide deep dives and insights for specific topics.

Keywords: Future technology, Smart coatings, Sustainability

POSTERS

Biobased, sustainable technologies

P-01 Novel bio-based polymer thermosets utilizing enzyme catalysis

Samer Nameer[1], Stefan Semlitsch[2], Mats Martinelle[2], Mats Johansson[1]
[1]Coating Technology, Stockholm, Sweden
[2]Industrial Biotechnology, KTH, Stockholm, Sweden

Introduction

Materials made from natural resources are regarded as sustainable and environmentally friendly. One example of such natural resource is biomass waste. Today biomass such as bark from the forest industry is regarded as a low value waste and is mainly used as heat production. One way to valorize biomass is to take advantage of interesting building blocks within the biomass. It has earlier been shown that birch bark contains interesting epoxide functional building blocks e.g. cis-9, 10-epoxy-18-hydroxyoctadecanoic acid (EFA) ([1], [2]). EFA can be extracted from birch bark by alkaline hydrolysis and can further be used for thermoset applications (2). In addition, it has earlier been shown that EFA can undergo enzyme catalyzed esterification reactions to create a homopolymer keeping the epoxy group still intact (3). Such procedure creates an opportunity to introduce interesting functionalities, such as thiols, to EFA and keeping the epoxy group intact. The aim of present studies is to utilize building blocks from biomass e.g. birch bark and synthesize thermosets for coating applications.

Results

In the present studies EFA extracted from birch bark has been used for thermoset applications. In a first study EFA showed an ability to cure into a thermoset with adhesive properties. The curing procedure was self-catalyzed by EFA at elevated temperatures. The adhesion ability of the bio-based thermoset was evaluated on different substrates. Moreover, in the second study the aim was to add thiol functionality to EFA by utilizing enzyme catalysis. The reaction containing an aliphatic thiol, EFA and dimethyl adipate was catalyzed by Candida Antarctica lipase B (CALB) in toluene. A thiol-epoxy functional polyester resin was obtained. The synthesized polyester resin provided a possibility to either react only the thiol or the epoxide selectively by choosing polymerization technique i.e. cationic polymerization or chain/ step wise polymerization. Furthermore it was possible to modify the surface properties of the coating by using one of the unreacted functionalities and react it with a desired molecule.

Conclusions

Biomass waste from the forest industry can be used to synthesize thermosets for coating applications. It was found that the natural occurring epoxy fatty acid (EFA) have adhesive properties. The thermoset was formed due to a self-catalyzed and self-condensation reaction at elevated temperatures. The carboxylic acid in the molecule acted as a catalyst hence, no external catalyst was needed to start the reaction. Furthermore, the enzyme CALB proved to be a perfect catalyst to add thiol functionality to epoxide containing monomer without harming the otherwise very reactive epoxide. Moreover, the synthesized thiol-epoxy polyester showed promising ability to cure into a coating that could be used to alter the surface properties of a substrate.

Acknowledgement

This work was supported by the Swedish research council, FORMAS (Grant No. 211-2013-70)

References:

1. Ekman R. The Suberin Monomers and Triterpenoids from the Outer Bark of Betula verrucosa Ehrh. Holzforschung-International Journal of the Biology, Chemistry, Physics and Technology of Wood. 1983;37(4):205-11.
2. Torron S, Semlitsch S, Martinelle M, Johansson M. Polymer Thermosets from Multifunctional Polyester Resins Based on Renewable Monomers. Macromolecular Chemistry and Physics. 2014;215(22):2198-206.
3. Olsson A, Lindström M, Iversen T. Lipase-catalyzed synthesis of an epoxy-functionalized polyester from the suberin monomer cis-9, 10-epoxy-18-hydroxyoctadecanoic acid. Biomacromolecules. 2007;8(2):757-60.

Keywords: Epoxy fatty acid, birch bark, CALB, coating

P-02 New concept for sustainable wood care products

Aaron Breivogel
Competence Center Lacquers & Varnishes, DAW SE, Ober-Ramstadt, Germany

Novel Sustainability Concept for Wood Care Products

The area of decorative and architectural coatings follows the worldwide trend towards new sustainable raw materials. “Sustainable” typically means “renewable”. However, renewable materials demand cropland area. Due to the demand of cropland area renewable materials are typically in competition with food production. Herein, we present a novel concept for sustainable raw materials which overcomes the competition with food production.

The concept does not compete with food production as existing pure pea fields are converted into mixed fruit fields of peas together with camelina. In this way, camelina is cultivated without using additional cropland. The goal is to create a highly sustainable supply chain, starting with the cultivation of camelina and ending in wood care products for varnishes and wood oils. On the one hand the farmers increase their financial income. On the other hand nature benefits from the positive effects on the ecosystem such as a stronger biodiversity and less pesticides on a large-scale area.

Peas and camelina plants are harvested in a single step before they are separated. While the farmers have a higher income due to the camelina crop, the amount of pea crop essentially stays the same.

An oil is extracted from the camelina seeds which is refined to a binder. The binder is part of wood care products (varnishes and wood oils) that enhance the lifetime e.g. of wooden fences or garden houses.

Peas and camelina develop synergies: camelina suppresses weed and therefore reduces the amount of necessary pesticides. The pea tendrils wind around the camelina plants which facilitates the pea harvest and increases the crop. Due to its rarity camelina is listed in the red list of endangered domestic crop plants in Germany. Mixed fruit fields of peas and camelina strengthen the biodiversity compared to pure pea fields. As long as camelina is in flower, it is an important food source for endangered insect species like pollinators that perform essential ecological system services.

The project is funded within the framework of Germany's federal program for biological diversity by the Federal Agency of Nature Conservation with resources from the Federal Ministry for the Environment, Nature Conservation, Building and Nuclear Safety.

The project consists of four work packages

1) Information of farmers:

Farmers are informed about the mixed fruit cultivation and are accompanied during the cultivation by an agricultural expert. Farmers get financial support: Full funding of the costs for the camelina seeds.

2) Creation of a Supply Chain and Optimization of the Conditions for Farmers:

Firms that separate the pea from the camelina crop are identified within the project. Oil mills and buyers for the filter cake are identified as well. Public funding for farmers is investigated. Farmers get full financial support in form of an insurance against vermins (vine fretter)

3) Public Relations:

Exhibitions, conferences, articles, online homepage

4) Scientific Research:

The effect on biodiversity of mixed fruit cultivation of peas together with camelina versus pure pea fields is scientifically investigated in cooperation with a university research institute.

Keywords: binder, Smart coatings, Sustainability, wood coating

P-03 The effect of surface pretreatments of mild steel on the film formation and corrosion resistance properties of vinyl silane coatings

Rahemeh Behgam, Mohammad Mahdavian
Surface Coatings & Corrosion, Institute for Color Science & Technology, Tehran, Iran

The proper surface pretreatment before coating application plays vital role in durability and any protective or even aesthetic functions of the coatings. Different surface pretreatment methods were used on mild steel substrate prior to application of vinyltrimethoxysilane (VTMS).
Electrochemical Impedance spectroscopy (EIS), polarization and electrochemical noise measurement (ENM) techniques were employed to study the corrosion protection properties of the coatings. In addition, surface morphology of the coating samples was examined using scanning electron microscopy (SEM).
The results showed that the coated sample pretreated by the means of titanat acidic solution had better film formation and corrosion resistance attributes.
The proper surface pretreatment before coating application plays vital role in durability and any protective or even aesthetic functions of the coatings. Different surface pretreatment methods were used on mild steel substrate prior to application of vinyltrimethoxysilane (VTMS).Electrochemical Impedance spectroscopy (EIS), polarization and electrochemical noise measurement (ENM) techniques were employed to study the corrosion protection properties of the coatings. In addition, surface morphology of the coating samples was examined using scanning electron microscopy (SEM). The results showed that the coated sample pretreated by the means of titanat acidic solution had better film formation and corrosion resistance attributes.

References:

[1] S. Bera, T.K. Rout, G. Udayabhanu, R. Narayan, Water-based & eco-friendly epoxy-silane hybrid coating for enhanced corrosion protection & adhesion on galvanized steel, Prog. Org. Coatings. 101 (2016) 24–44. doi:10.1016/j.porgcoat.2016.07.010.
[2] S. Alinejad, R. Naderi, M. Mahdavian, Effect of inhibition synergism of zinc chloride and 2-mercaptobenzoxzole on protective performance of an ecofriendly silane coating on mild steel, J. Ind. Eng. Chem. 48 (2017) 88–89. doi:10.1016/j.jiec.2016.12.024.
[3] M. Yu, M. Liang, J. Liu, S. Li, B. Xue, H. Zhao, Effect of chelating agent acetylacetone on corrosion protection properties of silane-zirconium sol-gel coatings, Appl. Surf. Sci. 363 (2016) 229–239. doi:10.1016/j.apsusc.2015.12.013.
[4] J. Wojciechowski, K. Szubert, R. Peipmann, M. Fritz, U. Schmidt, A. Bund, et al., Anti-corrosive properties of silane coatings deposited on anodised aluminium, Electrochim. Acta. 220 (2016) 1–10. doi:10.1016/j.electacta.2016.10.080.
[5] L. Lei, S. Jing, X. Wang, D. Liu, H. Xu, Microstructure and electrochemical behavior of cerium conversion coating modified with silane agent on magnesium substrates, Appl. Surf. Sci. 376 (2016) 161–171. doi:10.1002/sia.5734.
[6]E. Alibakhshi, E. Ghasemi, M. Mahdavian, B. Ramezanzadeh, S. Farashi, Fabrication and characterization of PO4 3− intercalated Zn-Al- layered double hydroxide nanocontainer, J. Electrochem. Soc. 163 (2016) C495–C505. doi:10.1149/2.1231608jes.
[7] S. Li, A. Pasc, V. Fierro, A. Celzard, Hollow carbon spheres, synthesis and applications – a review, J. Mater. Chem. A. 4 (2016) 12686–12713. doi:10.1039/C6TA03802F.
[8] J. Xiao, C. Zhao, C. Hu, J. Xi, S. Wang, Pudding-typed cobalt sulfides/nitrogen and sulfur dual-doped hollow carbon spheres as a highly efficient and stable oxygen reduction electrocatalyst, J. Power Sources. 348 (2017) 183–192. doi:10.1016/j.jpowsour.2017.03.011.

Keywords: corrosion resistance, electrochemical impedance spectroscopy, Silane coating, Surface Treatment

P-04 Effect of nano particles treated with silane coupling agent on physical and corrosion properties of heat shrinkable field joint coating

Babak Biuk Afshari[1], Mehran Rostami[2], Masoud Jamshidi[1], Iman Khaki[3]
[1]Chemical department, IUST University, Tehran, Iran
[2]Nano department, color and coating institude, Tehran, Iran
[3]SPD Co., Tehran, Iran

Protection with barrier effect is the main method for protecting the underground pipe line. The protection of this method completes with cathodic protection. So, the behavior of coatings when protect under the cathodic current is important. In pipe line, the field joint area is important for protection against corrosion. So the selection of field joint coatings must be according to the main coatings. In poly ethylene 3 layer coating system, for protection of field joint area, usually used heat shrinkable sleeve coatings. These coatings according to service temperature were classified. The aim of this study, assessment of surface modification of nanoceria and silica on physical and corrosion properties of hot melt adhesive for service temperature up to 60 °C. Surface treated nano particles 1, 3,5 % Wt. were added to hot melt adhesive. To study the chemical interaction of nano particles with polymeric matrix, using DSC, TGA and Softening point Analysis. Also to assessment electrochemical properties of adhesive, using EIS (Electrochemical Impedance Spectroscopy) and cathodic protection test.

P-05 Preparation of improved zinc-rich epoxy coatings by using polyaniline/graphene oxide nanopigments

Morteza Ebrahimi[1], Bahare Miri[2], Mohamad Reza Mohamadzade Attar[2], Farhad Sharif[2]
[1]Polyemer Eng. and Color Tech., Tehran, Iran
[2]Amirkabir University of Technology, Tehran, Iran

Zinc-rich epoxy primers (ZRP) are sacrificial coatings which protect steels from corrosion through cathodic protection. Zinc-rich epoxy primers require a very high content of zinc powder in formulation to protect the substrate properly. This very high content of zinc powder results in film porosity, loss of adhesion strength of coating to substrate and weakening of the mechanical properties of coating. High porosity in the film results in rapid consumption of active zinc powder and shortening the service life of the coating.
The aim of this study was to synthesis polyaniline-graphene oxide (Pani/GO) nanocomposites and use them in a zinc-rich formulation. In fact, it was reported that Pani/GO nanocomposites exhibited exceptional electrical conductivities and therefore, they can be replaced by a substantial amount of zinc powder in zinc-rich epoxy coatings. This can improve the corrosion resistance and physical-mechanical properties of the coatings at the same time which is economically and environmentally very beneficial. It should be noted that plate-like structure of GO could improve the barrier property of the coating as well. In this research 0.2, 0.5 and 1 wt.% of Pani/GO was replaced with 10 wt.% of zinc powder in a zinc-richepoxy coating formulation.
Electrochemical impedance spectroscopy (EIS) and open circuit potential (OCP) measurements of carbon steel coated panels showed that the use of Pani/GO nanocomposite in formulation improved barrier properties of coating compared with the reference coating. It was observed that the resistance of coating contained 0.2 wt.% nanocomposite with pigment volume concentration of 43 increased to 6860 ohms after 35 days of immersion in a 3.5% NaCl solution, while the resistance of reference zin-rich coating reduced to 207 ohms.
The reduction of zinc powder in zinc-rich epoxy coating improved the adhesion and mechanical properties of nanocomposite coatings compared to the reference zinc-rich epoxy coating.

References:

1. P. A. Sorensen, S. Kiil, K. Dam-Johansen, and C. E. Weinell, "Anticorrosive coatings: A review," J. Coatings Technol. Res., vol. 6, no. 2, pp. 135–176, 2009.
2. C. P. de Leo´n1, S. A. Campbell2, and J. R. Smith2 and F. C. Walsh, "Conducting polymer coatings in electrochemical technology Part 2 – Application areas," Trans. Inst. Met. Finish., vol. 86, pp. 34–40, 2008.
3. M. H. Mohamadzadeh Moghadam, S. Sabury, M. M. Gudarzi, and F. Sharif, "Graphene oxide-induced polymerization and crystallization to produce highly conductive polyaniline/graphene oxide composite," J. Polym. Sci. Part A Polym. Chem., vol. 52, no. 11, pp. 1545–1554, 2014.
4. E. Akbarinezhad, M. Ebrahimi, F. Sharif, and A. Ghanbarzadeh, "Evaluating protection performance of zinc rich epoxy paints modified with polyaniline and polyaniline-clay nanocomposite," Prog. Org. Coatings, vol. 77, no. 8, pp. 1299–1308, 2014.
5. H. Marchebois, S. Joiret, C. Savall, J. Bernard, and S. Touzain, "Characterization of zinc-rich powder coatings by EIS and Raman spectroscopy," Surf. Coatings Technol., vol. 157, no. 2–3, pp. 151–161, 2002.

Keywords: corrosion resistance, graphene oxide, Polyaniline, zinc-rich epoxy coatings

P-06 Effect of surface salt contaminations and curing content of fusion bonded epoxy (fbe) on interface porosity, pull of adhesion and cathodic disbonding

Iman Khaki[1], Mehran Rostami[1], Babak Biuk Afshari[2], Morteza Ganjaee[1]
[1]Nano department, Color and Coating Institude, Tehran, Iran
[2]Chemical department, IUST University, Tehran, Iran

Now days, for protection of underground pipe line, used special coatings with cathodic protection. Fusion bonded epoxy has been used in pipeline coating since 1980. In North America, FBE most commonly used coating for new pipeline construction. FBE coatings also used as the coating system for girth-weld. The quality of FBE coatings depending on curing content and pipe surface salt contamination. In this study, effect of epoxy curing value and surface salt contamination on electrochemical and physical properties were considered. To assessment corrosion properties, used electrochemical impedance spectroscopy (EIS) and cathodic disbanding tests. Also to consider the physical properties of FBE coating, used pull of adhesion and interface porosity tests.

P-07 Characterization of water and electrolyte transports in autophoretic corrosion protection coatings by means of optical spectroscopies and electrochemical impedance spectroscopy

Tobias Krawczyk[1], Michael Dornbusch[1], Jochen Gutmann[2]
[1]Hochule Niederrhein, Krefeld, Germany
[2]Universität Duisburg-Essen, Essen, Germany

Corrosion is a well-known issue in every region and every industry of the world. Despite it is hard to determine exactly it is estimated that in the industrialized countries corrosion destroys between 2.5% and 4.5% of the gross domestic product per year [1]. One possibility to prevent these damages is the use of organic corrosion protection coatings. If this coating is not damaged corrosion will be slowed down dramatically. But even an intact coating can be penetrated by water, electrolytes and other corrosion supporting substances which can cause corrosion under the coating.
Here the transport processes of water and electrolytes through a commercial available autophoretic coating system for steel is examined. The challenge is the pigmentation of the coating, because it makes it hard to receive valuable spectroscopic results. For this purpose, coated samples with varying pigment concentrations are treated with water or electrolyte solutions e.g. 3% NaCl in water for a defined period. After this the penetration depth of water and the electrolyte itself can be determined by means of different spectrometric methods like UV/Vis, IR, and Raman spectroscopy. Additional electrochemical impedance spectroscopy and scanning Kelvin probe gives another way to observe the samples water-uptake. All these methods in combination are used to receive a clearer image of the transport processes inside the bulk and at the coating/substrate interface and the coatings and in particular its pigment concentrations influence on it. Furthermore, the spectroscopic methods can be used to detect changes in the coatings structure or composition during the experiment. By analyzing the speed of water and electrolyte uptake with Fickian and Non-Fickian approaches a model for the transport of these corrosion supporting substances through a coating can be developed.
The correlation of coating composition, electrolyte transport through this coating, its structural changes and the observed corrosion processes can be analyzed. The received information is used for predicting favorable coating compositions and optimization of the resin/pigment ratio. This helps to create a coating with higher barrier effect against corrosion supporting substances, so that the corrosion is slowed down extremely.

References:

[1] Moulinier, F.; Techniques, Sciences, Methodes 7/8; 2010; 26-38

Keywords: Corrosion Protection, transport processes

P-08 The effect of Zn pigment modifications on the structure and mechanical properties of anticorrosive coatings

Ewa Langer[1], Małgorzata Zubielewicz[1], Helena Kuczyńska[1], Agnieszka Królikowska[2], Leszek Komorowski[2]
[1]Paint and Plastics Department, Institute for Engineering of Polymer Materials and Dyes, Gliwice, Warsaw, Poland
[2]Road and Bridge Research Institute, Warsaw, Poland

Typically, anticorrosive coating system consists of a primer, one or several intermediate coats, and a topcoat. The function of the primer is to ensure good adhesion to the substrate and to protect it from corrosion, mainly by the surface active compounds which are present there. For this purpose, zinc pigments are often formulated into such primers, where they act as sacrificial anodes. During lifetime of such systems Zn is transformed in its corrosion products, which tend to seal the pores between the Zn particles to a point at which the system becomes electrically nonconductive (barrier stage). Since the performance of sacrificial coatings is based on transfer of galvanic current, which implies metallic contact between the individual active pigments, such coatings are typically very highly pigmented, which sometimes causes losses in cohesive strength and in resistance to impacts. Therefore – and because Zn is known to be harmful to the environment and becoming a rare chemical element – research aims at Zn primers with reduced Zn content and even better long time performance. It may be pursued by use the mixture of Zn pigments with different shape (dust, flakes) and different size, Zn alloy pigments (Zn/Sn, Zn/M) as well as by chemical or metallic surface treatment of such pigments and an addition of graphene or carbon nanotubes.
Epoxy paint formulation with reduced amount of such Zn pigments were developed to test the influence of the concentration and shape, size and surface modification on the mechanical and anticorrosive properties of coatings. In this work the structure of coatings and their mechanical properties are presented in comparison with commercial Zn-rich primers. Testing of mechanical properties of coatings included flexibility, hardness, impact resistance and cupping test. Scanning electron microscopy (SEM) was used to evaluate the distribution of the pigments and to observe the contact between pigment particles and thermogravimetric analysis (TGA) to characterize coatings.
Test results showed some differences between properties of coatings depending on kind of modification of Zn pigments.

Acknowledgments
The research was carried out under the project CORNET/22/1/2016 "New generation of zinc primers with improved anticorrosion, application and ecological properties (ZINCPOWER)" funded by the National Centre for Research and Development (NCBiR).

Keywords: anticorrosive coatings, modification, Zn pigment

P-09 Rare earth la-modified $MoSi_2$ coating for oxidation protection of carbon/carbon composites

Kezhi Li
Northwesern Polytechnical University, Xi'an, China

Carbon/carbon (C/C) composites has attracted much attention due to its excellent mechanical properties at high temperature. However, it suffers easily from catastrophic oxidation in high temperature oxidizing environment, which limits its wide applications greatly. $MoSi_2$ is the most dependable coating material for refractory metals and alloys because it has an outstanding oxidation resistance up to a temperature of 1973K due to the formation of a self-healing, glassy silica (SiO_2) scale that can protect the substrate from oxidation and corrosion at elevated temperature. Disappointingly, some drawbacks for application of $MoSi_2$ at high temperature has been increasingly presented. For example, it undergoes pest disintegration around 773K and shows poor oxidation resistance between 673-1073K. The thermal stress induced by the coefficient of thermal expansion (CTE) mismatch between the coating and substrate is another issue that needs to be considered. When there exists the thermal stress, micro-cracks are easily formed.. Adding rare earth elements to $MoSi_2$, especially those who have larger affinity to oxygen than Si, is regarded as an effective way to enhance the high temperature performance of $MoSi_2$.

In the present work, lanthanum was designed as a rare earth element to produce a novel La-Mo-Si multiphase coating. The effectiveness of the designed coating in protecting C/C composites from oxidation at high temperature has been investigated. Cubic specimens (10mm×10mm×10mm), which are used as substrates, were cut from a bulk of C/C composites. They were hand-abraded by SiC paper, and cleaned with ethanol and dried at 353K. The powder compositions consisted of 5-20wt.% LaB6, 50-70wt.% $MoSi_2$ and 10-30wt.% SiC, which were dry-mixed and roller milled to ensure uniformity. The obtained mixed powders were agglomerated by a spray tower drier to enhance the fluidity of particles for spraying, and then sprayed to form the rare earth La-modified $MoSi_2$ coating with power of 50-55kW. The experimental results show that the designed coating system included a La-Mo-Si outer layer and a SiC transition layer. The rare earth lanthanum substitution for molybdenum changed the crystal structure of $MoSi_2$ from tetragonal C11b to C40, resulting in the formation of the dual-phase mixture containing C11b $MoSi_2$ and C40 (Mo, La) Si2. The dual-phase coating can provide a long-term protection for C/C composites at 1773K. Meanwhile, the introduction of lanthanum as the second phase to $MoSi_2$ resulted in the formation of a compact Si-La-O compound glass on the coating surface during oxidation, which is favorable to improve the oxidation resistance of the coating by restricting the inward diffusion of oxygen.

This work has been financially supported by the National Natural Science Foundation of China under Grant No.51432008.

Keywords: oxidation protection, $MoSi_2$ coating, carbon/carbon composites

P-10 Study on the electrical conductivity and anti-corrosion properties of epoxy coatings improved by graphene particles

Bin Liu
Materials Science and Engineering, Beijing University of Chemical Technology, China, Beijing, China

In this paper, graphite and graphene particles were used to reinforce the electrical conductivity and anti-corrosion properties of epoxy coatings for oil tank protection. The feature of graphite and graphene particles dispersed in the coating was characterized by scanning electron microscopy, transmission electron microscopy and Raman spectroscopy. At the same time, the electrical conductivity and protective properties of coatings were studied by surface resistivity tester and electrochemical impedance spectroscopy. The experimental results showed that epoxy coatings' performance could be enhanced by appropriate additive amount of graphene particles. The differential effect of graphite and graphene were compared and discussed.

Keywords: graphene, electrical conductivity, anti-corrosion, epoxy coatings

P-11 Diffusion of oh-solvents in organic coatings

Shicong Luo, Ting Wang, Claus E. Weinell, Søren Kiil, Kim Dam Johansen
Department of Chemical and Biochemical Engineering, Technical University of Denmark, Kgs. Lyngby, Denmark

Abstract
A one-chamber diffusion cell is used to measure the diffusion rate of methanol in organic coatings. The weight loss of the cell was intermittently measured on an analytical balance to estimate the diffusion coefficient of methanol. Typically, Fick's second law can be used to describe the diffusion process. However, due to the swelling of the organic coating layer and the sharp diffusion front phenomenon [1], a custom-made model should be developed for the system in the future.

Introduction
Organic protective coatings are widely applied to protect various metal surfaces such as ships, wind turbines, and oil rigs. These metal substrates are commonly, vulnerable to aggressive species such as acids or organic molecules. Therefore, a chemical resistance coating is used to protect underneath substrate from corrosion [2]. The basic protective mechanism for the coating system is the barrier effect. However, the penetration of corrosive chemicals limits the lifetime of the coating system [2,3]. Thus, it is important to study the diffusion of these chemicals to estimate the lifetime of organic coatings. In this study, a one-chamber diffusion cell is used to investigate the diffusion of methanol across organic coatings.

Experimental
A one-chamber diffusion cell is used to perform the diffusion experiments. The organic coating (Polyurethane (PU), solvent based) was placed between the aluminum cup and the flange. The aluminum cell was filled with methanol. The diffusion rate of methanol across the coating is determined by intermittently measuring the weight loss of the cell.

Mathematical model
The diffusion of methanol in the organic coating is generally regarded as concentration-dependent. Fick's second law, which is shown in equation (1), can be used to estimate the weight loss and diffusion rate during the experiment.

$$\partial C/\partial t=\partial[D(C)\cdot\partial C/\partial x]/\partial x \qquad (1)$$

where D(C) is the concentration-dependent diffusion coefficient, x is the distance, t is the diffusion time, and C is the penetrant concentration in the coating.
The average transmission rate of methanol (Q) can be defined as equation (2) [4]:

$$Q=D\cdot S \qquad (2)$$

where D is the average diffusion coefficient and S is the sorption coefficient.

Future work
Future diffusion experiment of methanol and butanol will be conducted using one-chamber cell. Additionally, in order to examine the applicability of Fick's second law to describe this system, mathematical fitting will be performed. However, if there is a relative large error between the experiment results and the mathematical model, a custom-made diffusion model of alcoholic solvents through the organic coating system is needed. The phenomena or reason for causing the error will also be studied.

References:
[1] G.S. Hartley, T. Faraday, Journal of the Chemical Society B, 1946, 42, 6–11.
[2] S. Romhild, G. Bergman, M. S. Hedenqvist, Journal of Applied Polymer Science, 2005, 95, 797-806.
[3] P. A. Sørensen, S. Kiil, K. Dam-Johansen, and C. E. Weinell, Journal of Coatings Technology and Research, 2009, 6, 135-176.
[4] M. Zhu, D. Vesely, European Polymer Journal, 2007, 43, 4503-4515.

Keywords: diffusion, mathematical model, polyurethane

P-12 Fabrication of an advanced anti-corrosion system based on epoxy/cerium oxide quantum dots on mild steel

Maryam Mousavi[1], Amir masoud Arabi[2], Bahram Ramezanzadeh[1], Mohammad Mahdavian[1]
[1]Department of Surface Coating and Corrosion, [2]Inorganic Pigment and Glazes Department, Institute for Color Science and Technology, Tehran, Iran

Cerium oxide quantum dots (CeQDs) were successfully synthesized and characterized by high resolution transmission electron microscopy (HR-TEM), field-emission scanning electron microscopy (FE-SEM) and UV-visible analysis. The optical band gap values were determined by energy wave equation and Tauc method. The corrosion inhibition effect of CeQDs in 3.5 wt.% chloride solution on mild steel was studied through electrochemical impedance spectroscopy (EIS). Subsequently the corrosion protection properties of mild steel samples coated with epoxy coatings loaded with CeQDs were investigated by electrochemical impedance spectroscopy (EIS) measurement and the salt spray exposure.
Results of FE-SEM, HR-TEM and UV-visible and band gap analysis confirmed that CeQDs with sizes less than Bohr radius of cerium oxide were successfully synthesized. The EIS test performed on the mild steel samples dipped in the solutions with CeQDs extract confirmed their active corrosion inhibition effect. Also, the enhancement of the corrosion protection properties of the epoxy coating filled by CeQDs was demonstrated by EIS and salt spray test results.

References:

1. Zhang, Z., Wang, L., Shao, M., Guo, Y., Chen, G., Shao, J., & Zhang, G. (2012). Synthesis andcharacterisation of CeO 2 nanoparticles and rods. IET Micro & Nano Letters, 7(8), 770-772.
Zhang, Z., Wang, L., Shao, M., Guo, Y., Chen, G., Shao, J., & Zhang, G. (2012). Synthesis and characterisation of CeO 2 nanoparticles and rods. IET Micro & Nano Letters, 7(8), 770-772.
2. Rai, P., Khan, R., Ko, K. J., Lee, J. H., & Yu, Y. T. (2014). CeO_2 *quantum dot functionalized ZnO nanorods photoanode for DSSC applications. Journal of Materials Science: Materials in Electronics, 25(7), 2872-2877.*
3.Arul, N. S., Mangalaraj, D., Chen, P. C., Ponpandian, N., & Viswanathan, C. (2011). Strong quantum confinement effect in nanocrystalline cerium oxide. Materials Letters, 65(17), 2635-2638.
4. Sahu, T., Singh Bisht, S., Ranjan Das, K., & Kerkar, S. (2013). Nanoceria: synthesis and biomedical applications. Current Nanoscience, 9(5), 588-593.
5. Ramasamy, V., & Vijayalakshmi, G. (2016). Synthesis and characterization of ceria quantum dots using effective surfactants. Materials Science in Semiconductor Processing, 42, 334-343.
6. He, X., Zhang, D., Li, H., Fang, J., & Shi, L. (2011). Shape and size effects of ceria nanoparticles on the impact strength of ceria/epoxy resin composites. Particuology, 9(1), 80-85.

Keywords: Cerium oxide quantum dots, Corrosion, Epoxy, Synthesis

P-13 Corrosion protection and heat resistance in a single coat

Tamas Pozman[3], Andrea Lukacs[1], Livia Molnar-Nagy[1], Ilona Felhosi[2]
[1]Research & Development, Industrial Paint Research Ltd., Budapest, Hungary
[2]Institute of Materials and Environmental Chemistry, Budapest, Hungary
[3]Technical service, Industrial Paint Research Ltd., Budapest, Hungary

The market demands that steel constructions (e.g. heaters, ovens, chimneys, furnaces BBQs) have to handle high temperatures and resist corrosion effects.
There are excellent solutions for high temperature resistance up to 500°C and for corrosion protection, but not many can combine these two criteria in case of a single layered coating.
Anticorrosive pigments are widely used in conventional coatings. Besides zinc phosphates, various ecologically favourable alternatives can be an option as a protective compound. Their performances are tested and approved at normal temperature conditions. Not many examples can be found about using them in coatings designed for steel constructions which operate at high temperatures. Therefore, at elevated temperature the anticorrosive compounds' behaviour and effectiveness are not well known.
In order to find an efficient solution, we started a series of examination which included raw material and complex matrix tests to find and approve a method where a coating can be thermoresistant and anticorrosive at the same time.
Salt spray and electrochemical impedance spectroscopy (EIS) were used to monitor electrochemical activity that occurs at the paint /substrate interface upon corrosion and degradation of metal to paint bonds. [1]
The aim of this study is to determine whether a single layered coating system with an anticorrosive pigment can provide a complex heat and corrosion protection on a conventionally used industrial carbon steel surface.

References:

[1] Jörg Sander, (2010) Anticorrosive Coatings, Vincentz Network, Hanover, Germany, ISBN 978-3-86630-911-1

Keywords: Anticorrosive pigment, Corrosion protection , Heat resistance, Polysiloxane coatings

P-14 Failure analysis of coating adhesion: peeling of internal oxidation layer over electrical steel after stress relief annealing

Hsin-Wei Lin
China Steel Corporation, Kaohsiung City, Taiwan, Republic of China

Electrical steel (ES) are commonly used in the production of power generators, motor cores and transformers. The internal stresses and strains of ES could be generated during manufacturing processes, therefore, need to be removed by stress relief annealing. However, several clients of China Steel Corporation reported that the coats were peeled-off after the heat treatment of ES. To resolve the problems and decrease the defect rate, this work focused on the analysis of peeled-off morphology and the optimization of heating parameters. Several instruments such as dual-beam focused ion beam, energy dispersive spectrometer, glow discharge spectrometer and transmission electron microscopy were used to characterize the microstructures of peeled-off and well-attached samples. The results showed that samples after heat treatment were composed of three layers of structure including coat, internal oxidation zone (IOZ) and substrate. Surprisingly, the occurrence of crack present in the interface between IOZ and substrate. When the IOZ was peeled from substrate, there is no possibility that the coats could still remain on the surface. Generally speaking, the peeled-off samples were featured as thick thickness of IOZ, in the meanwhile, containing high concentration of oxides in this region. With precisely control of gas composition and soaking time in the annealing section, the defect rate of adhesion failure could be effectively eliminated.

References:

1. W.F. Block and N. Jayaraman: J. Mater. Sci. Technol., 1986, vol. 2, pp. 22-27.
2. E. J. Hilinski and G. H. Johnston: "Annealing of electrical steel" in Electric Drives Production Conference, Nuremberg, Germany, Sept. 30-Oct. 1, 2014.
3. J.S. Park and J.T. Park: "Effect of stress relief annealing temperature and atmosphere on the microstructure and magnetic properties of non-oriented electrical steels" in Electric Drives Production Conference, Nuremberg, Germany, Nov. 30-Dec. 1, 2016.
4. G. Lyudkovsky, A. G. Preban and J. M. Shapiro: J. Appl. Phys., 1982, vol. 53, pp. 2419-2421.
5. S.C. Paolinelli and M.A. da Cunha: J. Magn. Magn. Mater., 2006, vol. 304, pp. e599-e601.
6. J.C. WU, P.C. SUN and P.L. CHEN: China Steel Technical Report, 2008, No. 21, pp. 38-44.
7. K.Y. PENG, K.C. HSIEH, R.M. CHEN and Y.P. SUI: China Steel Technical Report, 2012, No. 25, pp. 42-50.
8. C.K. KUO and J.C. WU: China Steel Technical Report, 2010, No. 23, pp. 46-51.
9. K. Yanagihara and S. Yamazaki: Nippon steel technical report, 2011, No. 100, pp. 27-32.
10. L.L. Liu, Q.Q. Guo, S. Liu, C.S. Ni and Y. Niu: Corros. Sci., 2015, vol. 98, pp. 507-515.
11. W.F. Block and N. Jayaraman: J. Mater. Sci. Technol., 1986, vol. 2, pp. 22-27.
12. E. J. Hilinski and G. H. Johnston: "Annealing of electrical steel" in Electric Drives Production Conference, Nuremberg, Germany, Sept. 30-Oct. 1, 2014.
13. J.S. Park and J.T. Park: "Effect of stress relief annealing temperature and atmosphere on the microstructure and magnetic properties of non-oriented electrical steels" in Electric Drives Production Conference, Nuremberg, Germany, Nov. 30-Dec. 1, 2016.
14. G. Lyudkovsky, A. G. Preban and J. M. Shapiro: J. Appl. Phys., 1982, vol. 53, pp. 2419-2421.
15. S.C. Paolinelli and M.A. da Cunha: J. Magn. Magn. Mater., 2006, vol. 304, pp. e599-e601.
16. J.C. WU, P.C. SUN and P.L. CHEN: China Steel Technical Report, 2008, No. 21, pp. 38-44.
17. K.Y. PENG, K.C. HSIEH, R.M. CHEN and Y.P. SUI: China Steel Technical Report, 2012, No. 25, pp. 42-50.
18. C.K. KUO and J.C. WU: China Steel Technical Report, 2010, No. 23, pp. 46-51.
19. K. Yanagihara and S. Yamazaki: Nippon steel technical report, 2011, No. 100, pp. 27-32.
20. L.L. Liu, Q.Q. Guo, S. Liu, C.S. Ni and Y. Niu: Corros. Sci., 2015, vol. 98, pp. 507-515.

P-15 New polyurethane-based lubricious coatings: tuning the network architecture and hydrophobic-hydrophilic balance towards a low friction surface

S.P.W. Govers[1], P.T.M. Albers[1], L.G.J. van der Ven[1], R.A.T.M. van Benthem[1, 2], G. de With[1], A.C.C. Esteves[1]
[1]Laboratory of Physical Chemistry, Department of Chemical Engineering and Chemistry, Eindhoven University of Technology, Eindhoven, the Netherlands
[2]DSM Ahead BV Netherlands, Geleen, the Netherlands

Lubricious coatings are widely applied in the biomedical sector, being an essential element of many medical devices. These coatings generally exhibit a low friction coefficient by making use of a highly hydrated polymer network.[1, 2] The exceptional properties of such systems, e.g. their low elastic modulus and high water content, provide systems with potentially unique surface low-friction behavior.[3] However, their mechanical properties are often suboptimal.4 To improve these characteristics, new types of polymer network coatings were investigated, where hydrophilic polymer chains are introduced into a densely crosslinked hydrophobic polymer network. In water, these hydrophilic chains can form a thin, highly hydrated surface layer providing excellent lubricating properties.
Polyurethane (PU) network coatings were prepared by crosslinking poly(propylene glycol) (PPG) using a tri-isocyanate crosslinker and incorporating poly(ethylene glycol) methyl ether (mPEG) as a hydrophilic dangling chain or poly(ethylene glycol) (PEG) as a hydrophilic network chain component. By varying the PPG and (m)PEG precursor molecular mass and ratio in the network, a wide range of different PUPPG-(m)PEG systems has been synthesized. The friction behavior of these systems was investigated using a home-made tribology setup with a glass countersurface. The lubricating properties of PUPPG network coatings were significantly improved upon introduction of hydrophilic polymer chains in the network, both as a network chain (PEG) or as a dangling chain (mPEG). These coatings were able to achieve a coefficient of friction (CoF) of 0.01 without the high bulk swelling and low elastic modulus often found for comparable systems. Low surface friction coatings with improved mechanical properties could be successfully realized by tuning the polymer network architecture and balance between hydrophobic and hydrophilic components in the network.

References:

1. *S. Nagaoka and R. Akashi, Biomaterials, 1990, 11, 419-424.*
2. *P. Wyman, in Coatings for Biomedical Applications, 2012, pp. 3-42.*
3. *J. P. Gong, Soft Matter, 2006, 2, 544-552.*
4. *M. L. Oyen, International Materials Reviews, 2013, 59, 44-59.*

Keywords: Lubrication, Polyurethane network coating

P-17 Microstructure and wear-resistance performance of laser hybrid plasma spraying tic-nicraly coatings

Shuqing Li, Qilian Li, Guojia Ma
Beijng Aeronautical Manufacturing Technology Research Institute, Beijing, China

The TiC-NiCrAlY coatings are deposited on 38CrMoAl substrate by the laser-hybrid plasma spraying (LHPS) technology in this study. Test results showed that the LHPS coating's microstructure became more dense, its bonding strength was greatly increased and micro-hardness were well distributed when the NiCrAlY content accounted for 70%(30%TiC-70%NiCrAlY). Then the 30%TiC-70%NiCrAlY coating was also prepared by the air plasma spraying (APS) technology in order to compare with the LHPS coating. SRV high-temperature friction and wear tests revealed that, when compared with as-sprayed ceramic coatings, the wear rate is greatly reduced due to the addition of the laser power in LHPS. During the LHPS processing, the laser irradiation melt the spraying powders sufficiently and densified the coating's microstructures, and coating's porosity was decreased greatly. Therefore, there were no obvious cracks and cavities in the coating. The main wear mechanism of the as-sprayed coatings was spallation of the coating and abrasive wear, whereas it was dominated by ploughing and abrasive wear for the LHPS specimens.

Keywords: coating, Laser hybrid plasma spraying, TiC-NiCrAlY, wear-resistance performance

P-18 Slippery paints: eco-friendly coatings to which crawling insects cannot adhere

Aurélie Féat[1], [2], Phil Taylor[1], Marleen Kamperman[2], Walter Federle[3], Jasper van der Gucht[2]
[1]AkzoNobel Decorative Coatings, Slough, United Kingdom
[2]Physical Chemistry and Soft Matter, Wageningen University, Wageningen, the Netherlands
[3]Department of Zoology, University of Cambridge, Cambridge, United Kingdom

Controlling insect pests in buildings is important for reducing disease transmission and structural damage (termite damage alone is estimated at $40 billion p.a. [1]). To minimize the use of potentially harmful insecticides [2], new environmentally friendly strategies explore physical rather than chemical mechanisms to control insects.
Many natural plant surfaces (e.g. pitcher plants [3]) reduce or prevent insect adhesion via (1) multi-scale surface roughness, (2) lubricating layers, and (3) wax crystals detaching from the surface.
The main goal of this project is to formulate environmentally friendly paints, which provide an anti-adhesive mechanical barrier for crawling insects on vertical surfaces.
Model paint systems containing only polymer binder, pigment, filler and water have been prepared. It was found that controlling the following parameters can strongly decrease insect adhesion to paint surfaces: (1) the type, diameter (< 5 µm) and geometry of paint pigment particles; and (2) the chemistry, particle size (> 100 nm) and glass transition temperature (> 20°C) of the binder polymers, as shown in preliminary studies.
Results show that the solid particles at the top surface often detached and adhered to the tacky pads of insects [4]. In optimized formulations, 85% to 100% of tested Atta cephalotes ants tested slipped and fell from vertical paint panels.
After 2000 cycles of cleaning with an abrasive solution, paints have shown good durability and were still highly slippery as pigment particles detached more easily from the coating surface. Rather than surface parameters (micro-roughness, surface energy), the results suggest that particle detachment is the main mechanism which reduces insect adhesion due to feet contamination.
These paints are hence potential durable and environmentally friendlier alternatives for insecticidal paints and insecticides in buildings.

References:

[1] G. K. Mahapatro, D. Chatterjee. P. Natl. A. Sci. India B, 1-18, 2017
[2] G. J. Devine, M. J. Furlong. Agric. Human Values, 24, 3, 281–306, 2007
[3] E. Gorb, K. Haas, A. Henrich, S. Enders, N. Barbakadze, S. Gorb. J. Exp. Biol., 208, 4651-4662, 2005
[4] M. J. Anyon, M. J. Orchard, D. M. A. Buzza, S. Humphries, M. M. Kohonen. J. Exp. Biol., 215, 605-616

Keywords: Contamination, Eco-friendly coating, Insect adhesion

P-19 The use of the sol gel-synthesized borosilicate glass powder as a hierarchical surface element in the hydrophobic modification of the glass surface

Yiğitalp Okumuş[1], Göktuğ Günkaya[2]
[1]Graduate School of Sciences, Department of Materials Science and Engineering, Eskişehir, Turkey
[2]Faculty of Fine Arts, Department of Glass, Anadolu University, Eskişehir, Turkey

Hydrophobicity of micro, nano scaled hierarchical surfaces with lowered surface energy has been studied by today's researchers. In this present study, it was aimed to form a hierarchical surface by using glass powders having small particle sizes in the coating solution which was applied to glass surfaces for reducing the surface energy. Glass powders were synthesized by sol-gel method at room temperature in the presence of ammonium hydroxide as an alkali catalyst and ethanol-water mixture as solvent. Average particle size of synthesized glass powders was 60 nm. For the purpose of obtaining hydrophobic property on the borosilicate substrates, glass powders were used as additives in the chemical modification material to form rough surfaces. Two solutions were prepared by using different silane types and five different amounts of glass powders were added to each. Borosilicate glass substrates were coated with the prepared coating suspension. The coated samples were heat treated for curing of silane matrix at 150°C for an hour and then washed with methanol and water. The thicknesses of the resulting coatings, the morphology and the size of the particles were visualized by scanning electron microscopy using secondary electron detector. Contact angle measurements were carried out on static mode to determine the changes of water contact angle values and consequent hydrophobic features.

Keywords: Coatings on Glass, Contact Angle, Hydrophobic Surfaces, Sol-Gel Glasses

P-20 Study on hydrophobic surface formation on metal substrate using alumina pastes

Seong-Dae Park, Hyunseung Yang, Min-Jeong Na, Hye-Jin Jung
Korea Electronics Technology Institute, Seongnam-si, South Korea

Hydrophobic and superhydrophobic coatings are very attractive technologies which are generally used in diverse industrial fields. As an example, a heat exchanger like low pressure feedwater heater or condenser in power plant uses many of pipes in which cooling water flows. To improve the efficiency of heat exchanging, the hydrophobic coating can be formed on these pipes. The reason is that this hydrophobic coating induces dropwise condensation of steam instead of filmwise condensation on metal pipe surface.

We studied the method for hydrophobic coating structure formation on metal surface using Al_2O_3 paste, alkyl amines, and organic-inorganic hybrid coating sols. From the view point that hydrophobic effect can be determined by surface roughness and chemical composition on solid surface, we tried to form surface roughness on metal surface by Al_2O_3 paste and use coating sols with low surface energy which gives high hydrophobicity.

Materials used for this study are like these: α-Al2O3(4N, powder, ca.1μ, KOJUDNO KOREA CO. LTD), Glass frit(SF-2700, BASS), Ethyl cellulose, Alkyl amines, PDMS-based organic-inorganic hybrid sol, Ethylene glycol butyl ether, and 1-dodecanol. Alumina pastes with different alumina/glass ratios were fabricated using alumina powders, glass frit and ethyl cellulose-based vehicle. PDMS-based organic-inorganic hybrid sol was synthesized as described in the previous report. Measured water contact angle of this sol was about 100°. Alumina paste was screen-printed on SUS substrate, dried, and then fired at 500 °C in box furnace. Next, the PDMS hybrid sol or the alkyl amines were coated on this alumina surface by dip coating or spray coating and then heat-treated. From this methods, we investigated the hydrophobic effect according to the alumina/glass weight ratio and the kinds of coated solutions.

Surface roughness and micromorphology of the alumina paste-printed surface were evaluated. We could find that the higher surface roughness was shown with increasing alumina/glass weight ratio: Ra~0.666um at 30/55, 1.050um at 50/35, and 2.151um at 80/5. The surface images by FE-SEM also confirmed this roughness difference depending on the compositional changes. Contact anlges of hydrophic surfaces formed on SUS plates were measured after application and heat treatment of alkyl amines or PDMS coating sol. They were increased from 24° to 120° with increasing alkyl chain length of alkyl amine and increasing amine concentration in solution. The contact angles of PDMS-based sol-coated surfaces were varied from 100° to 136° depending on the change of alumina/glass ratio. As the conclusions, we could find the effects of surface roughness formed by the alumina pastes and coating solutions on hydrophobic efficiency, or contact angle, of the coated structure. In the future, we have a plan to improve the heat resistance and the durability of this coated structure based on the alumina paste.

Acknowlegment: This work was supported by the Korea Institute of Energy Technology Evaluation and Planning(KETEP) and the Ministry of Trade, Industry & Energy(MOTIE) of the Republic of Korea (No.20152020105710).

Keywords: alumina paste, contact angle, hydrophobicity, surface roughness

P-21 Quantification of leveling in conventional anti-fouling coatings

Yixin Huang[1], Søren Kiil[1], Xueting Wang[1], Claus E. Weinell[1], Stefan Møller Olsen[2]
[1]Chemical and Biochemical Engineering, Technical University of Denmark, Lyngby, Denmark
[2]Hempel A/S, Lundtoftegårdsvej 91, DK-2800, Lyngby, Denmark

Abstract
Leveling is a crucial phenomenon for coating surface structure after application, which is influenced by coating properties and application condition [1]. In this study, a novel method for evaluating leveling is presented using a 3D profilometer combined with an automatic film applicator. Based on the obtained roughness and waviness profiles of coating surface structure, a theoretical model to describe the effects of various factors on surface leveling phenomenon is formulated through experimental investigation.

Introduction
For conventional anti-fouling coatings, the surface structure of coating film plays an important role in coating performance owing to its influence on friction drag, anti-fouling properties and aesthetic appearance [2]. To optimize coating surface structure, it is crucial to investigate leveling which is a surface tension-driven phenomenon in drying process. In literature, some theoretical equations have been derived to describe this phenomenon [3]. However, certain assumptions were made for those models, which is not applicable for conventional anti-fouling coatings. The aim of this project is to study the effects of various physical parameters on leveling and to generate a new theoretical equation to quantify leveling process for conventional anti-fouling coatings via experimental investigations.

Experimental methodology
A model formula of conventional anti-fouling coatings will be used in this work. This project will focus on several influencing parameters, e.g. film thickness, wavelength and viscosity. The main equipment is the combination of an automatic coating applicator with a 3D profilometer. It can perform coating film application with provoking sinusoidal surface structure and surface observation instantly. In this method, human operating error can be avoided during coating film application. The surface structure is recorded every 15s. After applying various filters with different cut-off lengths according to International Organization for Standardization (ISO), the roughness and waviness profiles will be generated, and the relevant roughness and waviness parameters can be calculated and analyzed using software.

Future experimental work
Firstly, the influence of variables (thickness, wavelength and viscosity) on leveling will be determined individually as a function of flash off time. Besides, during a drying process, the changes of wavelength, film thickness will be observed by the 3D profilometer. The viscosity change can be estimated by knowing the changes of the solid content. Subsequently, the changing curves of film thickness, wavelength and viscosity versus drying time will be obtained. Meanwhile, the roughness and waviness profile changes will be recorded using the equipment as well. Thus, a mathematical model to describe the influence of each parameter along time on leveling will be formulated.

References:
[1] M. P. Schultz, Biofouling, 2007, 23, 331-341.
[2] S. E. Orchard, Applied Scientific Research, Section A, 1963, 11, 451-464.
[3] F. Seeler, C. Hager, O. Tiedje, &M. Schneider, Journal of Coatings Technology and Research, 2017, 14, 767-781.

Keywords: Anti-fouling coatings, Leveling quantification, Surface structure

P-22 Polyurethane coatings based on star polymers prepared by ionic polymerization

Peter Bohacik[1], Stepan Podzimek[2], [3], Vladimir Spacek[3], Jan Banzduch[3], Michaela Kadlecova[2]
[1]Institute of Chemistry and Technology of Macromolecular Substances, Pardubice, Czech Republic
[2]University of Pardubice, Pardubice, Czech Republic
[3]Synpo a.s, Pardubice, Czech Republic

This work is focused on the preparation of star polymers by ionic polymerization. Star polymers with different arm numbers were prepared from different ratio of arm forming chemicals. Trimethylol propane, Pentaerythritol and Dipentaerythrythol were used as core forming chemicals. Ɛ-Caprolactame and lactic acid were used for arm forming with different ratio of both. The yeld of all synthesis was about 99 %. The prepared polymers were characterized by Size Exclusion Chromatography coupled with static Multi Angle Light Scattering (SEC-MALS).

The prepared stars were then combined with aliphatic diol (linear polyol was prepared by Group Transfer Polymerization, and grafted by Ɛ-Caprolactame and isophorone diisocyanate (IPDI).
The crosslinked films were evaluated by tests widely used in paint industry, e.g. pendulum hardness, specular gloss, microhardness, determination of stone-chip resistance of coatings. Films were tested by Differential Scanning Calorimetry (DSC) to determine of glass transmission (T_g), and selected coating were tested by Dynamic Mechanical Analysis (DMA) for investigation of Young´s modulus of elasticity.

We found increase of hardness with increasing amount of aliphatic diol in the all prepared coatings. This effect we attribute to decreasing amount of star polymer, which acts like plasticizer. The coating has high gloss, good resistance against methyl ethyl ketone (MEK test). In comparison coating based on clear graft polyol cured by IPDI has greater determination of stone-chip resistance of coatings, resistance against abrasion and high gloss. The T_g is increases, this effect is cost by increasing amount of polyol in the coating. Stars in the coatings have plasticizing properties.

This coating based on pure polyol will be further formulated for e.g. in aircraft industry, automotive industry.

Keywords: polyurethane coating, SEC-MALS, star polymers

P-23 The influence of primary/secondary hydroxy group content in acrylic polymer on the curing and performance of 2k-PUR coatings

Aleksei Kostitsyn[1], Irina Grianko[1, 2], Viktoriia Soloveva[1], Aleksandr Ilin[2], Evgenii Indeikin[1, 2]
[1]YarLI - Yaroslavl Paint Research Institute, Yaroslavl, Russian Federation
[2]Yaroslavl State Technical University, Yaroslavl, Russian Federation

2K polyurethane coatings are now widely used in industrial coating applications. Acrylic styrene copolymers are often used as hydroxy-functional polymers. Depending on the hydroxy-functional co-monomers applied (e.g., hydroxyethyl methacrylate or hydroxypropyl methacrylate), the copolymer will contain primary or secondary hydroxy groups. The position of hydroxy group in a polymer backbone must influence the rate of its reaction with isocyanate cross-linking agent and, consequently, the coating material and film properties.

In the present study acrylic styrene copolymers were used that differed in hydroxy group type (primary, secondary or both). The amount of hydroxy-functional monomer was chosen in such a way that the hydroxy group content in the copolymer was within the range of 2,5-3%. The amount of styrene was the same in all the polymers, i.e. 42 %. All of them had approximately the same glass transition temperature (35-38 °C), which was achieved by changing the methyl methacrylate / butyl acrylate ratio. Isocyanate hardener - HDI-biuret – was used as a cross-linking agent.

Using small-amplitude oscillatory shear (SAOS) technique and DMA it was demonstrated that the rate of cross-linked polymer matrix formation differs, depending on the hydroxy group type. Moreover, the cross-linking density depends both on the hydroxy group type and on the curing temperature. With the increase of the curing temperature the zone of diffusion control of the cross-linking reaction becomes wider and so the network density and, as a consequence, the protective performance are improved.

The addition of secondary hydroxy groups into the acrylic oligomer leads to a slight decrease in the polymer matrix cross-linking density, whereby this change becomes more pronounced with the decrease of the curing temperature. At the same time, the addition of secondary hydroxy groups increases the mixture pot-life before application, which is a very important application parameter.

It was demonstrated that it is possible to control both the material properties before application and the protective performance of the coating film by changing the primary / secondary hydroxy group ratio in the acrylic copolymer.

P-24 The formation of polyester-melamine / urethane coatings

Nataliia Skopintseva[1], Nadezda Denisova[1, 2], Daria Afanaseva[1], Evgenii Indeikin[1, 2]
[1]YarLI - Yaroslavl Paint Research Institute, Yaroslavl, Russian Federation
[2]Yaroslavl State Technical University, Yaroslavl, Russian Federation

Polyester-melamine systems have become widely used in coil coating. They constitute the major part of the products applied in this industry. In spite of their great popularity and a high amount of previous studies the task of their improvement remains a priority. There are a high number of options to influence the process of fast thermal curing and the formed film performance. These are: the chemical composition of the polyester resin, the amount and the chemical composition of the cross-linking agents, the nature and the quantitative composition of the solvent mixture, the chemistry and the amount of the catalysts. One of the cost-effective ways to significantly influence the coating formation process and performance is to use a composite cross-linking agent containing an isocyanate. It can be expected that the presence of polyurethane links formed in a three-dimensional polymer coating film adhering to the substrate can provide a good balance of properties, such as hardness and flexibility, chemical resistance and weatherability, decorative properties, with the economic benefits of their application in comparison to pure polyurethane products.

In our investigations a hybrid cross-linking agent was used for curing a saturated polyester with a hydroxyl value of 120 mgKOH/g and an acid value of 3 mgKOH/g. Hexa(methoxymethyl)melamine and methyl ethyl ketoxime-blocked hexamethylene diisocyanate were used as components of the hybrid cross-linking agent. Blocked p-toluenesulfonic acid and dibutyltin dilaurat, respectively, were used as catalysts.

Based on DMA and DSC measurements it was demonstrated that the reaction between the saturated polyester and hexa(methoxymethyl)melamine starts at a lower temperature than that between the polyester and the isocyanate. This is, first of all, due to the difference in the deblocking temperatures of the catalysts. Thus when the two above mentioned cross-linking agents are used in a combination the main polymer matrix is formed by reaction of the polyester with hexa(methoxymethyl)melamine. Diisocyanate reacts with the sterically available rests of the unreacted hydroxy groups. It can also react with low molecular weight substances that are formed by co-condensation of the polyester with hexa(methoxymethyl)melamine, in particular, with methanol. The process of thermal curing of the composition also involves partial self-condensation of hexa(methoxymethyl)melamine.

Synchronization of chemical processes that determine the polymer matrix homogeneity of the composite coating on the substrate can be achieved by variation of the catalyst system composition, the cross-linking agent concentration ratio, the solvent mixture composition and the temperature of the coating formation process.

P-25 Application of layered double hydroxide nanocontainers as smart host of corrosion inhibitors into a silane/ epoxy bilayer coating

Eiman Alibakhshi, Akram LabbaniMotlagh, Ebrahim Ghasemi, Mohamamd Mahdavian, Bahram Ramezanzadeh
Institute for color science and technology, Tehran, Iran

Layered double hydroxide (LDH) nanoparticles are used as nanocontainers for reserving corrosion inhibitive compounds and releasing them in corrosion environment through an ion exchanges process. In this paper, the effect of inclusion of layered double hydroxide (LDH) nanoconatainers, as smart host of corrosion inhibitors, into a silane/epoxy bilayer coating was investigated. For this purpose, Zn-Al LDH with nitrate as the charge balancing anion in the interlayer gallery was prepared using a co-precipitation method. Then, the Zn-Al-NO_3^- LDH nanoparticles were pillared with phosphate and molybedate ions through anion-exchange reaction. The nanocontainers were characterized by X-ray diffraction (XRD), Fourier transform infrared (FT-IR) spectroscopy and scanning electron microscopy (SEM). Then, the active corrosion protection properties of the LDHs into a silane/epoxy bilayer coating were studied by electrochemical impedance spectroscopy (EIS) and electrochemical noise (EN) techniques. In this paper, a multi-resolution wavelet analysis was performed for feature extraction from the electrochemical signals. In this regard, the electrochemical signals are decomposed in predefined number of scales (or crystals) in undecimated wavelet transform (UWT) framework. EIS and EN results revealed that the bilayer coatings in the presence of phosphate could provide better corrosion protection and self-healing properties compared to the other corrosion inhibitors.

References:

1. *Electrochimica Acta, 2012, 82, 314-323.*
2. *Surface and Coatings Technology, 2014, 258, 17-37.*
3. *Corrosion Science, 2010, 52, 602-611.*
4. *Corrosion Science, 2017, 115, 159-174.*

P-26 Improvement in corrosion resistance of mild steel by applying polymeric coating incorporating halloysite loaded with an in-organic inhibitor

Najmeh Asadi[1], Reza Naderi[1], Mohammad Mahdavian[2]
[1]Metallurgy and materials engineering, University of Tehran, Theran, Iran
[2]Surface Coating and Corrosion Department, Institute for Color Science and Technology, Tehran, Iran

Aluminosilicate halloysite nanotubes (HNT) with a 15 nm lumen, 50 nm external diameter and length of 300 nm to 1 μm have been recently used as a container for storage and sustained release of anti-corrosion agents and drug delivery systems.
In this purpose, halloysite nanotubes loaded with an inorganic corrosion inhibitor are embedded into a polymeric coating applied on mild steel substrate to improve corrosion resistance of the metallic substrate. Release of the inhibitor from halloysite nanotubes have been proven by ICP-OES. In comparison with the blank coating, electrochemical impedance spectroscopy showed superiority of the
coating loaded with loaded halloysite particles.

References:

1) *D. Fix, D.V.A., Y.M. Lvov, D. G. Shchukin, H. Mohwald, Application of Inhibitor-Loaded Halloysite Nanotubes in Active Anti-Corrosive Coatings. advanced functional materials, 2009. 19: p. 1720-1727.*
2) *Yuri Lvov, H.M., Halloysite Clay Nanotubes for Controlled Release of Protective Agents. acs nano, 2008. 2(5): p. 814–820.*
3) *Peng Yuan, D.T., Faïza Annabi-Bergaya, Properties and applications of halloysite nanotubes: recent research advances and future prospects. Applied Clay Science, 2015. 112-113: p. 75-93.*
4) *Yuri Lvov, E.A., Functional polymer–clay nanotube composites with sustained release of chemical agents. Progress in Polymer Science, 2013. 38: p. 1690-1719.*
5) *V. Palanivel, Y.H., W. J. van Ooij, Effects of addition of corrosion inhibitors to silane films on the performance of AA2024-T3 in a 0.5 M NaCl solution",. Progress in Organic Coating, 2005. 3: p. 153.*
6) *K. A. Yasakau, M.L.Z., O. V. Karavai, M. G. S. Ferreira, Influence of inhibitor addition on the corrosion protection performance of sol–gel coatings on AA2024. Progress in organic coating, 2008. 63: p. 352.*
7) *Elshad Abdullayev, V.A., Asel Tursunbayeva, Vasiliy Portnov, Hikmat Ibrahimov, Gulbaniz Mukhtarova, Yuri Lvov, Self-Healing Coatings Based on Halloysite Clay Polymer Composites for Protection of Copper Alloys. ACS Applied Material Interfaces, 2013. 5: p. 4464-4471.*

Keywords: Halloysite, nanotubes, polymeric coating, electrochemical impedance spectroscopy

P-27 Molybdenum disulfide (MoS_2) nanoparticles application in active protective epoxy coating on mild steel

Seyyedarash Haddadi[1], Majed Amini[1], Ahmad Ramazani S.A.[1], Mohammad Mahdavian[2]
[1]Chemical & Petroleum Engineering Department, Sharif University of Technology, Tehran, Iran
[2]Surface Coating and Corrosion Department, Institute for Color Science and Technology, Tehran, Iran

The objective of this study is to develop and characterize the active protection properties of the zinc (Zn^{2+}) cations intercalated MoS_2 nanoparticles in an epoxy coating. The samples were prepared by direct dispersion of unmodified and zinc intercalated MoS_2 in various contents in epoxy resin. Inductively coupled plasma-optical emission spectrometer (ICP-OES) was employed to characterize ion exchange capability of intercalated MoS_2 in 3.5 wt. % NaCl solution. The active corrosion protection properties of the coatings were evaluated using electrochemical impedance spectroscopy (EIS), electrochemical noise measurement (ENM) and salt spray test. ICP-OES results revealed that the Na^+ cations presented in the galleries of MoS_2 were successfully exchanged with Zn^{2+}) cations. EIS, ENM and salt spray results showed that the coatings containing modified MoS_2 nanoparticles have higher anticorrosion performance in comparison with both blank and unmodified MoS_2 filled coatings.

P-28 Organosoluble polyimides and aromatic polyamides as the optical fibre coatings

Bato Bayminov[1], Dmitry Sapozhnikov[1], Nadezhda Popova[1], Sergey Semjonov[2], Alexey Kosolapov[2], Yakov Vygodskii[1]
[1]Russian Academy of Sciences, A.N. Nesmeyanov Institute of Organoelement Compounds, Russian Federation
[2]Russian Academy of Sciences, Fiber Optics Research Center, Moscow, Russian Federation

For high-performance optical fibers exploited in aggressive environment conditions and high temperatures of up to 300°C and above (oil and gas wells, sensors in spacecraft) special protective polymer coatings are needed [1]. One approach to solve this problem is to use protective polyimides (PI)coatings. Commonly the formation of PI coating is based on the use of hydrolytically and thermally unstable precursors, such as poly (amic acid)s (PAAs), which undergo imidization reaction upon coating formulation. This is accompanied by a number of drawbacks including PAA incomplete cyclization, a few solution deposition cycles, storage of PAA solution at reduced temperatures and so on. Moreover, a majority of the known PI's coatings has insufficient adhesion to the quartz fibre. The aim of this work is the elaboration of new types of organosoluble polyimide and polyamide coatings, which will eliminate the difficulties mentioned above.
We present the results on the fabrication of new optical fibre coatings from cardo (co)polyimides [(co)PI] and (co) polyamides [(co)PA] differed in improved heat resistance with organosolubility [2].
The obtained polymers have a wide range of molecular weights (η_{inh} =0.3÷1.8 dl/g) and thermal characteristics: 285°Cg<390°C; 420°C 10%<565°C (in air). Cast polymers films have high tensile strength (85 ÷ 140 MPa) and elasticity modulus of (1100 ÷ 2800 MPa).
The optical fiber coatings were produced from the varnishes of synthesized polymers in N-methyl-2-pyrrolidone (N-MP) with or without a coupling agent. It was established the optimum dynamic viscosity range necessary for the formation of a high-quality coating is 2.4 ÷ 18 Pa × s.
The as-prepared fibres with smooth polyimide coatings without any visible heterogeneity or flaws have high bending strength (~6 GPa) with small scatter [3]. The produced optical fibers are characterized by high thermal stability: their bending strength after holding at 300°C (72 hrs) and 350°C (1 h) is more than 90 % from initial one. It was observed the developed polyimide coatings exceed the commercial product (Ceram Optec) by the hydrolytic stability. Fabricated fibres with smooth (co)PA coatings demonstrated the similar tough and thermal stability.
Significant advantages over the existing technology are the possible coating formation by one deposition cycle with needed thickness (~ 5 microns), disposal of chemical transformation during the coating fabrication and as a consequence the absence of defective PAA fragments, there is no need to store the polymer solutions at low temperatures.
This work is supported by the Ministry of Education and Science of the Russian Federation (Unique identifier of Applied Scientific Researches (project) RFMEFI60417X0183).

References:

1. K. Sohma, T. Hattori, Proc. 60th Intern. Wire & Cable Symp. (Charlotte, NC, USA, 2011, p. 424).
2. S.L. Semjonov, D.A. Sapozhnikov, D.Yu. Erin, O.N. Zabegaeva, I.A. Kushtavkina, K.N. Nishchev, Ya.S.Vygodskii, E.M. Dianov. Quantum Electronics. 45:4 (2015), 330.
3. W. Weibull, J. Appl. Mech.-Trans. ASME 18:3 (1951), 293.

P-29 In-line dispersion techniques for efficient coating production for efficient coating production

August Boye[1], Søren Kiil[1], Claus E. Weinell[1], Kim Dam-Johansen[1], Przemyslaw Debinski[2]
[1]Department of Chemical and Biochemical Engineering, Lyngby, Denmark Technical University of Denmark
[2]Hempel A/S, Kgs. Lyngby, Denmark

Background

Coatings have traditionally been produced in batch-mode, with the dispersion of pigments taking place first. However, recent in-line dispersion principles have made it possible to make coating production much more efficient. Currently, the in-line mixers have shown very promising results for many pigments, but require more work to be an established technique for all types of pigment agglomerates.

The dispersion process consists of three steps. First, the pigment agglomerates all have to be wetted. Secondly, the agglomerates have to be broken down to provide the desired fineness of grind. Lastly, these primary particles have to be stabilized, so that the pigments do not re-agglomerate or precipitate in the solution. How these steps are achieved depends on the specific equipment used. Conventional dispersion equipment relies on generating high shear forces to break the agglomerates. Newer methods mix the pigments under vacuum to break them using a pressure difference.

Aims

The aim of this project is to characterize the efficiency of in-line mixers compared to conventional batch dispersion techniques.

Methods

First, the literature on coating production and dispersion theory has been studied and described to understand the relevant working principles. Based on this the in-line mixer and conventional dispersion techniques were characterized and compared in terms of achieved fineness of grind, particle size distribution and stability. The fineness of grind and pigment particle size distribution have been measured with laser diffraction and used to compare the two production methods. Various pigment types and binder systems were investigated. Lastly, the dispersion kinetics of selected pigments in both dispersers were also analyzed.

Results

The dispersion of pigments in coatings has been found important for coating properties such as color strength, optical and application properties. This can be in both the degree (size and distribution) and stability of the dispersion. The dispersion process is also one of the most expensive steps, especially for coatings requiring high dispersion [1]. Traditionally, high-speed dispersers produce high shear forces using a rotating disc to break the agglomerates apart. Milling includes several forces: rotational, centrifugal, gravity, and lifting, and largely depend on the adhesive and cohesive nature of the system [1]. The in-line mixer uses a different principle, by introducing the pigment under vacuum. This subsequently creates a pressure gradient between the core of the agglomerate and the surrounding media, which forces the solvent/binder system into the pores of the agglomerates, wetting all the surfaces and breaking the agglomerate. This technique shows great promise, at is avoids the intense mechanical forces required by conventional dispersion methods.

References:

[1] L. Lin, (2003) "Mechanisms of pigment dispersion", Pigment & Resin Technology, Vol. 32 Issue: 2, pp.78-88

Keywords: Dispersion techniques, In-line mixing, Semi-continuous production

P-30 New method for objective hiding assessment of wall paint

Steven De Backer
Titanium Technologies, Chemours, Kallo, Belgium

The intrinsic hiding power of a paint is very well described by the theory of Kubelka and Munk, resulting in what is often called a theoretical spread rate. This number expresses the number of square meter of wall that can be covered at a certain contrast ratio. The test method assumes a very uniform application, which can be realized under lab conditions, but does not necessarily corresponds to the hiding power of a paint layer applied with a roller. Such a roller creates, pending on the flow characteristics of the paint, the substrate, the type of roller,... a more or less pronounced surface structure (peaks and valleys), resulting in a reduction of the observed hiding power. In this paper we describe how this surface structure can be quantified via statistical image analysis of the paint surface after roller application on a black substrate.

Keywords: applied hiding power, flow, intrinsic hiding power, roller application

P-31 Easy-to-disperse-fumed silica: efficient way to reduce production costs

Marco-Yann Heuer [1], Guenther Michael[2]
[1]AT Industrial Coatings EMEA, EVONIK Resource Efficiency GmbH, Essen, Germany
[2]Product Development, EVONIK Resource Efficiency GmbH, Hanau, Germany

For rheology control in coatings fumed silica is one of the state-of-the-art products. Due to progress in the last years to create easy dispersable colorants and pigments also investigations of easy-to-dispers fumed silica for rheology control took place. As an outcome of these research tests a range of products was developed for faster and easier production. Instead of the necessarily used bead or pearl mills with high shear forces only the usage of a dissolver is needed to achieve identically rheological results in coatings. So it is possible to speed up the production of coatings significantly by using only a dissolver. Even a post-addition at the end of the process is possible, which was in the past only an option with the addition of special manufactured intermediate goods.
The production process of creating a special particle selection is able to create properties with the new easy-to-dispers silica rheology modifiers like conventional fumed silica rheology modifiers.
Particle distribution of the new products simulates the values like the treatment of the conventional products after treating them with a pearl or ball mill.

Keywords: efficency, Production, rheology, silica

P-32 Dispersion using the vacuum expansion method

Hans-Joachim Jacob
Ystral GmbH, Ballrechten-Dottingen, Germany

Complete wetting and agglomerate free dispersion of every single pigment or filler particle is the pre-condition of a high quality paint or coating. Traditional High speed Dissolvers require a massive amount of wetting agent. The amount is often 50 to 5.000 times higher than acceptable in the finished paint. The result: foam, the need for defoamers and many negative side effects of the surfactants in the finished coating surface.

The newest Inline Disperser Technology has started to displace the High Speed Dissolvers worldwide. It uses a totally different principle: the controlled in-line High Shear Dispersion in combination with the Vacuum Expansion Method.

The vacuum expansion principle separates particles already on their way into the liquid before they are wetted. The particles are wetted and dispersed completely within fractions of a second. Even agglomerates are destroyed completely under shear and the vacuum expansion effect.

The dispersion conditions are completely different from High speed dissolvers. Wetting agents are not required. Defoamers are eliminated or at least reduced as well. This leads to raw material savings and a higher quality of the finished coating.

Dispersions with particle sizes below 1 μm down to below 100 nm are reached. 90% of the pigment pastes produced with this method do not need any milling process anymore. Time- and cost savings up to 90% are often reported. Paint- and ink production gets much more effective.

But even new functional layers based on nanotechnology with characteristics like easy-to-clean, anti-graffiti, scratch resistance, elasticity and transparency are produced this way very effectively.

Keywords: Dispersion, Milling, Pigment, surfactant

P-33 In-line quality control for efficient coating production

Sigurd Bay Jørgensen[1], Claus Erik Weinell[1], Kim Dam-Johansen[1], Henrik Bach Falkenberg[2], Søren Kiil[1]
[1]Chemical and Biochemical Engineering, Technical University of Denmark, Lyngby, Denmark
[2]Hempel A/S, Kgs. Lyngby, Denmark

Introduction

Whether it is weighing of raw materials, the dispersion of the millbase, the grinding in the pearl mill, the mixing in of the let-down or the final adjustment of colour and viscosity, most steps in current coating production are performed one batch at a time with thourough cleaning and quality control in between. This is a time consuming process leading to underutilized equipment, requiring larger stock and more plant personnel. Additionally, the coatings industry is challenging like never before with increasing complexity in regulation and pressure on raw material prices. Therefore, for products produced in large volume (>200,000 liters per year in a single factory), the incentive to convert to efficient in-line dispersion techniques in coating production is massive.

This transition to new production principles requires implementation of in-line quality control, which can replace the traditional labour and time-intensive methods. Two instruments capable of on-line measurement of very important parameters, fineness of grind and viscosity, have been identified: Insitec/Mastersizer from Malvern Instruments, which use laser diffraction to measure the particle size distribution of a sample [1], and Rheostream from Fluidan, which can measure various rheological parameters through pressure drop data [2].

Previous work

Previous work on in-line quality control of coating production has already been performed at the Technical University of Denmark [3]. It was found that the Insitec/Mastersizer was more precise at measuring the fineness of grind compared to the typical grindometer and that solvent shock was not a problem. No testing on the Rheostream was performed in the previous work, but it was argued that it could also be used to estimate the sagging index. It was emphasized that both equipment require calibration for each new formula, however.

Aims of the work

The goal of this project is to further improve on the work previously performed at the Technical University of Denmark with regards to implementation of in-line quality control in production of coatings. A preliminary plan set out is:

- Design of modular-based in-line dispersion techniques in coating production
- Mapping of relevant quality control methods
- Selection of 1-2 quality control methods and experimental investigation using them
- Benchmarking against existing quality control methods
- Practical recommendation

References:

[1] Malvern Instruments. Mastersizer 3000. Retrieved from https://www.malvern.com/en/assets/MRK1872_tcm22-17232.pdf
[2] Fluidan. (2017). Company presentation
[3] Palazón, L. B. (2016). Continuous Production of Solvent-Based Coatings. Technical University of Denmark, Master Thesis.

Keywords: Control, In-line, Production, Quality

P-34 Development of model system for intelligent screening of conversion coatings

Polina Volovitch, Thomas Sanchez, Simon Gillet
Chimie ParisTech, Paris, France

Correct choice of conversion coating (CC) which promotes paint adhesion to a metallic substrate and improve anticorrosion and antiblistering resistance of the whole painted system is essential for lasting the life time of part of painted metallic materials [1]. The most universal conversion coatings formulations known since decencies were based on the Cr(VI) [2]. Numerous formulations exempt of Cr(VI) have been developed since more than 10 years, however their action seems to be much more sensible to the composition and microstructure details of the substrate than it was the case of hexavalent chromium based CC. The choice of an appropriate formulation for a selected substrate and in particular for an assembly becomes hence a complex task.
Systematic testing of available CC formulations in real industrial systems is often time consuming, expensive and the obtained ranking cannot give an insight into the failure mechanisms at buried metal-conversion coating –polymer interface. In particular, high electric resistance, high thickness and complex chemistry of industrial polymers makes difficult to access the evolution of the interface in situ, to localize the week points and to study degradation mechanisms in detail by electrochemical methods or using surface analysis. Several mechanistic in-situ studies of degradation at metal-polymer interfaces using Scanning Kelvin Probe, Time Lapse Microscopy or Local Electrochemical Impedance Mapping were made with polyvinilbutiral (PVB) and related week polymers [3-5]. The adhesion of PVB to metal substrates is however too week and PVB is not stable in water for long times, different mechanisms of deadhesion should be hence expected under this polymer and under more realistic organic coatings. As a result, PVB seems not to be an ideal material for in situ studies of CC materials in which interface degradation is supposed to be long. Similarly, it cannot be considered for immersion studies aiming to understand the role of CC for blistering.
The development of model systems with higher degradation rate than in case of industrial organic coating but still enough resistant to immersion in water, with lower electrical resistance than industrial polymers but still offering some barrier effect in order to support similar reactivity mechanisms at the interface and transparent to visible light in order to offer a possibility to survey the degradation of the interface in situ could help to understand these degradation mechanisms and advance an intelligent screening of conversion coatings for a selected substrate. In the present study we select criteria for a model polymer for in situ studies of metal-conversion coating-polymer interface for Zn based coatings, propose a methodology to characterize the stability of a selected resin on the example of diglycidylether bisphenol A triethylene tetramine (DGEBA TETA) epoxy resin using infrared spectroscopy, contact angle measurement and electrochemical impedance spectroscopy. We will illustrate the examples of application of this material for in-situ discrimination of surfaces which behaved very similar in electrochemical tests without polymer.

References:

[1] G.D. Wilcox, D.E. Walke. In: Q. J. Wang, Y.W. Chung (eds), Encyclopedia of Tribology. Springer, Boston, MA (2013)
[2] P. L. Hagans, C. M. Haas. In: C.M. Cotell, J.A. Sprague, and F.A. Smidt, Jr. (eds), ASM Handbook, Volume 5: Surface Engineering, ASM International (1994)
[3] G. Williams and H. N. McMurray, J. Electrochem. Soc. 148 (2001) B377-B385
[4] M. Rohwerder, S. Isik-Uppenkamp, C.A. Amarnath, Electrochim. Acta 56 (2011) 1889–1893
[5] V. Shkirskiy, P. Volovitch, V. Vivier, Electrochim. Acta 235 (2017) 442-452

Keywords: conversion coatings, degradation, Epoxy, interface

P-35 Small-scale fire tests for hydrocarbon intumescent coatings

Ying Zeng[1], Claus Erik Weinell[1], Kim Dam-Johansen[1], Louise Ring[2], Søren Kiil[1]
[1]Chemical Engineering, Technical University of Denmark, Kgs. Lyngby, Denmark
[2]Hempel A/S Fire Protection-Group R&D, Polinyà, Spain

The protection of structural steel on exposure to a fire has become paramount to prevent the loss of lives and assets with increasing use of structural steel as a material to create the shape of construction projects. Structural steel loses bearing ability when its temperature exceeds 500 °C (failure temperature), and an efficient way to prolong the time before the failure temperature is reached is to use intumescent coatings. At elevated temperature, intumescent coatings swell, via a complex chemical mechanism, to a porous multicellular char layer that acts as a thermal shield protecting the underlying substrate. For the special case of hydrocarbon fires on oil rigs, so-called hydrocarbon intumescent coatings are used, which must also be able to withstand jet fires after expansion.

The crucial assessment of intumescent coatings is fire testing that is usually done with big industrial furnaces under standard temperature-time curves. However, the industrial fire tests have shortcomings of high costs, poor repeatability, and longtime cooling demand between two tests. These factors limit standard fire tests, especially for hydrocarbon fires. Therefore, a small-scale and high-throughput fire testing is badly needed to favor the assessment and development of effective intumescent coatings.

The fire testing that we have developed is based on a lab-scale oven that can be rapidly heated up to 1600 °C within 10 minutes. The standard fire curve mimicked by this oven is UL 1709, which is a widely accepted hydrocarbon fire-testing curve for intumescent coatings. The samples for tests are prepared by applying about 6 mm intumescent coating on the surface of a grit-blasted steel plate. A thermocouple is attached to the back of the plate to record the temperature-time (T-t) relationship during the fire testing.

According to the chamber temperature recorded during each fire test, this oven can adequately simulate the T-t curve defined in the standard. The relative standard deviations of the time for the steel plate to reach 400 and 550 °C are less than 1%. Seven intumescent coatings will be used for experimentation in the lab oven and in industrial furnaces to link their performance and verify credibility of the lab oven for the fire tests under standard fire curves. Two coatings have presently been evaluated and the results indicate a high consistency between the lab oven and the industrial furnace with respect to the temperature- time relationship. This suggests that the lab oven is a promising small-scale solution for the standard fire tests. Additional experiments will be conducted to investigate the influence of coating ingredients on the intumescence performance and shed light on the intumescence mechanisms.

Keywords: fire protection, hydrocarbon intumescent coatings, standard fire test

P-36 Study of molecular weight development in emulsion polymerization of of styrene-acrylate and methylmethacrylate-acrylate copolymers

Roman Valka, Stepan Podzimek, Jana Machotova
Department of Chemistry and Technology of Macromolecular Substances, University of Pardubice, Pardubice, Czech Republic

The work deals with the preparation of acrylic copolymers via emulsion radical polymerization and study of molecular weight development of styrene n-buthylacrylate; styrene-ethylacrylate; methylmethacrylate- n-buthylacrylate and methylmethacrylate-ethylacrylate copolymyers (with different weight ratio of monomers) determined by asymmetric flow field-flow fractionation (A4F) with multi-angle light scattering detector (MALS). The polymerizations were carried out in water at 85° C using ammonium persulfate as initiator. The samples for determination of molecular weight were taken during polymerization at given time. The solid content of samples and conversion of the monomers was also investigated. The solid content of prepared latexes was about 30% and overall conversion of monomers was about 97%. It was found that the molecular weigh decreased with increasing content of methylmethacrylate (MMA) and styrene (ST) comonomers in the recipe. The influence of ST content on the molecular weight was greater then the influence of MMA content. The addition of 5 wt% ST led to an important reduction of molecular weight. In the case of MMA, a similar decrease of molecular weight was achieved by the addition of 25 – 50 wt % MMA. This finding can be explained by steric bulk of styrene units attached to the polymer chain that block the backbone tertiary C-H bonds which are the centers of transfer reactions, thereby preventing transfer reactions to the polymer.

Keywords: emulsion polymerization,A4F - MALS, molecular weight, chain reaction to polymer,

P-37 Conformational study of melamine crosslinkers and spectroscopical comparison of hmmm molecules by practical measurements and quantum chemical calculations

Janek Wysoglad[1, 2], Jan-Eric Ehlers[2], Michael Dornbusch[3], Jochen S. Gutmann[4]
[1]Department of Chemistry, University of Duisburg-Essen, Essen, Germany
[2]BU Coating BBA-Technologie Color, thyssenkrupp Steel Europe AG, Kreuztal, Germany
[3]Department of Chemistry and Institute for Coating and Surface Technology, Niederrhein University of Applied Sciences, Krefeld, Germany
[4]Physical Chemistry & CENIDE, Department of Chemistry, University of Duisburg-Essen, Essen, Germany

To understand and improve network formation processes and performance properties makes structural molecular studies critically important for the coil-coating industry. Due to the high molar masses and predominantly terminal OH-groups in polyester binder systems for coil coating processes it is hardly possible to crosslink pure polyester molecules. Therefore, crosslinking agents such as hexamethoxymethylmelamine (HMMM) or less methylated derivatives, e.g. methoxymethylmelamine (MMM), are often added to the coating formulation [1]. For molecules with considerably fewer atoms and accordingly less rotational freedoms (such as MMM) it is quite possible to appreciate essential conformations. Thus, an MMM conformer study is straightforward and serves as orientation and oversimplification concerning HMMM conformer studies. For HMMM molecules we developed an extensive computation method to investigate the conformational distribution average and probably the most likely molecular structures. Using the density functional theory (DFT) B3LYP method with Dunning's correlation basis set, calculations were performed to investigate the three-dimensional structural geometries of HMMM (basis set cc-pVDZ) and MMM (basis set cc-pVTZ). Beginning with 1500 conformations for HMMM and using various cut-off filters we focused on final residual 22 conformers for solvent phase calculations and 16 conformers for gas phase calculations. To the best of our knowledge, this is the first time that chemical properties for melamine crosslinkers were presented under consideration of conformational population distribution. Thus, computations of full optimized structural geometries, energies and vibrational states indicate that preferred structural alignments for the methoxymethyl (MM) group in melamine molecules exist. Nevertheless, we suppose that as a matter of principle the MM group of MMM molecules can perform rotations by itself easily and all conformational structure geometries of MMM will exist. Concerning the MM groups position comparing to triazine plane HMMM molecules have two main configurations, "3up-3down" and "4up-2down", as dominant conformer species. Computations of weighted and averaged IR and Raman spectra of final conformations for HMMM monomers, dimers and trimers at ambient temperature are novel and in good agreement with experiment. Prediction of final UV-Vis spectrum for top ten Boltzmann-weight and averaged monomeric HMMM conformers is in perfect agreement with practical measurement of commercially available HMMM. Finally, using the optimized structural geometries for melamine molecules we open up new possibilities to simulate crosslinking reactions using quantum chemical calculations.

References:

[1] W.R. Zhang, T.T. Zhu, R. Smith, C. Lowe, An investigation on the melamine self-condensation in polyester/melamine organic coating, Prog. Org. Coat. 69 (4) (2010) 376–383

Keywords: computational chemistry, conformational analysis, melamine crosslinker, Raman, infrared, UV-Vis, spectroscopy

P-38 Synthesis and characterization of anticorrosive hybrid epoxy ester coatings containing polyaniline/ nanodiamond nanostructures

Rahman Mohammadkhani* [1], Akbar Shojaei[1], Nahid Pirhady Tavandashti[2]
[1]Chemical and Petroleum Engineering, Sharif University of Technology, Tehran, Iran
[2]Department of Materials Science and Engineering, Tehran Science and Research Branch, Islamic Azad University, Tehran, Iran

Abstract
In this study, hybrid polyaniline (PANI)/nanodiamond nanostructures containing various percentages of amino-functionalized nanodiamond (ND-NH2) were synthesized via a chemical route. At first, NDs were carboxylated at 420°C under the air atmosphere in a furnace. Thionyl chloride was used in order to chlorinate the carboxylated NDs. Then, ethylenediamine was added as a surface modifier to synthesize the amino-functionalized nanodiamond. After that, in-situ polymerization of polyaniline (emeraldine salt) was performed in presence of surface modified NDs to prepare hybrid polyaniline/nanodiamond nanostructures. Different morphologies of polyaniline/nanodiamond nanostructures from nanofibers to nanospheres were obtained by changing the concentration of NDs. The nanostructures were characterized by Fourier Transform Infrared Spectroscopy (FTIR), UV-visible spectroscopy, and Field Emission Scanning Electron Microscopy (FESEM). In order to prepare anticorrosion coatings, 0, 0.3, 0.6 and 1 wt.% of spherical form of polyaniline/ ND were introduced into epoxy ester resin, applied to carbon steel substrates. Corrosion protection performance and the adhesion of the fabricated coatings to the metallic substrates were evaluated by using Electrochemical Impedance Spectroscopy (EIS), salt spray, and pull-off tests, respectively. The results showed that the epoxy ester coatings containing hybrid polyaniline/nanodiamond nanostructures have superior corrosion protection performance compared to the blank epoxy ester coatings.

Keywords: Anticorrosion, Hybrid Nanostructure, Nanodiamond, Polyaniline

P-39 Performance SiC coating deposited on AISI 304 stainless steel exposed to the phenomenon of cavitation erosion

Vite-Torres Manuel, Villanueva-Zavala Arturo, Moreno-Ríos Marisa, Ezequiel A. Gallardo-Hernández, Farfán-Cabrera Leonardo I., Zaragoza-Granados Jonathan
SEPI-ESIME-Campus Zacatenco, Instituto Politécnico Nacional, Ciudad de México, Mexico

In this paper was analysed of the performance of the coating of SiC deposited on AISI 304 stainless steel under erosion cavitation. The coating SiC was deposited via Chemical Vapour Deposition (CVD). The testing was made in a vibratory apparatus ultra-sonic at frequency of 28 KHz. The experiments were initially conducted in tap water. Afterwards, a solution of tap water with 5 % abrasive particles of size of 20μm of aluminum oxide (Al_2O_3) to accelerate wear damage on specimen surface and replicate more aggressive erosive-cavitation scenery. The temperature of the media test was between 22 – 28° C. The wear progression was evaluated by measuring the mass loss in the specimens. Also the features on the scars wear were analysed where the wear mechanisms present were also identified, using optical microscopy and SEM. The wear progression was evaluated by measuring the mass loss in the specimens. According to the results, in both media the SiC coated specimens were more resistant to the cavitation erosion than the uncoated specimens

References:

[1] Cheng, F., & Jiang, S. (2014). Cavitation erosion resistance of diamond-like carbon coating on stainless steel. Applied Surface Science, 292, 16–26. http://doi.org/10.1016/j.apsusc.2013.11.044

[2] Chi, S., Park, J., & Shon, M. (2015). Study on cavitation erosion resistance and surface topologies of various coating materials used in shipbuilding industry. Journal of Industrial and Engineering Chemistry, 26, 384–389. http://doi.org/10.1016/j.jiec.2014.12.013

[3] Qiu, N., Wang, L., Wu, S., & Likhachev, D. S. (2015). Research on cavitation erosion and wear resistance performance of coatings. Engineering Failure Analysis, 55, 208–223. http://doi.org/10.1016/j.engfailanal.2015.06.003

[4] Qin, C. P., Zheng, Y. G., & Wei, R. (2010). Cavitation erosion behavior of nanocomposite Ti-Si-C-N and Ti/Ti-Si-C-N coatings deposited on 2Cr13 stainless steel using a Plasma Enhanced Magnetron Sputtering process. Surface and Coatings Technology, 204(21-22), 3530–3538. http://doi.org/10.1016/j.surfcoat.2010.04.012

[5] Hu, H. X., Zheng, Y. G., & Qin, C. P. (2010). Comparison of Inconel 625 and Inconel 600 in resistance to cavitation erosion and jet impingement erosion. Nuclear Engineering and Design, 240(10), 2721–2730. http://doi.org/10.1016/j.nucengdes.2010.07.021

[6] Li, Z., Han, J., Lu, J., Zhou, J., & Chen, J. (2014). Vibratory cavitation erosion behavior of AISI 304 stainless steel in water at elevated temperatures. Wear, 321, 33–37. http://doi.org/10.1016/j.wear.2014.09.012

[7] Laguna-Camacho, J. R., Lewis, R., Vite-Torres, M., & Méndez-Méndez, J. V. (2013). A study of cavitation erosion on engineering materials. Wear, 301(1-2), 467–476. http://doi.org/10.1016/j.wear.2012.11.026

[8] Pierson H. O., Hanbook of Chemical Vapor Deposition (CVD), Principles, Technology, and Aplications, Norwich, New York, U.S.A: WILLIAM ANDREW PUBLISHING, LLC, 1999

[9] Lasorsa C., P. Pineda Ramos y J. I. Volosín, "Recubrimiento duro de carburo de silicio sobre substratos metálicos y de silicio aplicados mediante la técnica de plasma CVD" Gcia. Invest. Y Aplic. Nucleares. Comisión Nacional de Energía Atómica. Argentina.

[10] Martínez Arriero Z., Master's thesis "Análisis del desgaste por erosión por impacto de partículas sólidas en recubrimientos de Carburo de Silicio (SiC) y Carbono Tipo Diamante (DLC) sobre sustratos de acero inoxidable" IPN-SEPI ESIME –UZ, México, D. F. 2015.

[11] Bregliozzi, G., Di Schino, A., Ahmed, S. I. U., Kenny, J. M., & Haefke, H. (2005). Cavitation wear behaviour of austenitic stainless steels with different grain sizes. Wear, 258(1-4 SPEC. ISS.), 503–510. http://doi.org/10.1016/j.wear.2004.03.024

[12] ASTM standard G32-06 , Standard Test Method for Cavitation Erosion Using Vibratory Apparatus, PA 19428-2959, USA,(2016).

Keywords: Cavitation erosion, Cavitometer Tester, SiC coating, Tribology

P-40 Evaluation of dispersion of calcium carbonate in an acrylic paint film using image processing

Parisa Najmi, Navid Keshmiri, Mohsen Mohseni, Saeideh Gorji
Polymer Engineering & Color Technology, Amirkabir University of Technology, Tehran, Iran

The degree of particle dispersion affect many properties of the liquid paints as well as dry films such as mechanical, rheological and optical. There are numerous methods to evaluate dispersion which are more or less qualitative, so finding a simple but precise method in this scope is essential. In the present study image processing techniques were applied to quantify the dispersion of calcium carbonate in acrylic paint films. To this end, the filler at different concentrations and in the presence and absence of dispersing agent was dispersed in the binder. An optical microscope was utilized to capture the dry film images. Three different methods such as box-counting, CR (contrast ratio) and histogram distribution of the particles were studied. CR method failed to obtain desired and accurate results. In box-counting each image was divided into sub-images. The percentages of particles (N) in each sub-image computed and standard deviations of N values were considered as a criterion of dispersion. By box-counting and histogram distribution methods could evaluate the dispersion up to specific PVC and until the images include pleasant information from the surface. Latter methods are responsive for evaluating the dispersion of particles in binder up to PVC = 15%.

Keywords: Acrylic Paint Film, Calcium Carbonate, Dispersion, Image Processing

P-41 Ferro functional pigments solution for coatings: deeper knowlegde of Nubirox anticorrosive pigments by the use of different techniques

Patricia Perez[1], Esther Oro[2]
[1]Corrosion Inhibitors Product & Technical Service Manager, L'Hospitalet de Llobregat, Spain
[2]Corrosion Inhibitors R&D Manager, Ferro Performance Pigments Spain, L'Hospitalet de Llobregat, Spain

Corrosion is recognized as a major problem in modern society and the resulting losses each year are in the hundreds of billions of dollars across many industries. Therefore, it is really a key issue to protect metallic surfaces so as to slow down and minimize the negative impact of corrosion. Coating the surface with an anticorrosive paint is one of the most common ways to achieve this target. And, in addition to the barrier effect of the coating, corrosion protection can be enhanced in many cases by the use of anticorrosive pigments inside the coating.

Ferro's range of corrosion inhibitors includes technically efficient anticorrosive pigments (Nubirox) that help coatings formulators to achieve present performance requirements for many protective coatings applications. In addition to their effectiveness in slowing down the corrosion of different metal surfaces, Nubirox are also an environmentally friendly solution for new sustainable coatings, as well as a compatible solution for innovative DTM coatings with additional aesthetics and gloss requirements.

The study of the activity and the behavior of the different particle shapes, chemistries and modifications involved in our Nubirox anticorrosive pigment range has helped us to clarify their activity, their performance inside a coating and their benefits in comparison with conventional pigments:

- Pigment particle shape (characterized by electronic microscopy, SEM) and its specific surface area (measured by BET) were analyzed as a reference of the pigment's active area, which correlates with its anticorrosive effectiveness.
- The inhibition mechanism of anticorrosive pigments is based on substrate passivation, and it was identified by using electrochemical characterizations (potentiodynamic polarization of a pigment water extract, LSV)
- But the anticorrosive pigment performance inside a coating not only depends on the pigment activity; coating formulation and the anticorrosive pigment interaction with other paint components are also important points to consider.

The prediction of anticorrosive properties in coatings is commonly evaluated by international standards like Salt Spray test (ISO 9227, ASTM B117), but modern electrochemical techniques are becoming increasingly useful to obtain additional information about the corrosion process in a short time. In particular, the "Accelerated Cyclic Electrochemical Technique" ACET (ISO17463:2014) was used to study the resistance of the coating to degradation (with cathodic polarizations), besides the common Salt Spray test. The results obtained show how the anticorrosive pigment effectiveness is also reflected in this technique.

The conclusion of the research carried out using these different techniques demonstrates the high effectiveness of our current Nubirox range, and it is part of our fundamentals for the development of future anticorrosive pigments.

Keywords: Anticorrosive coating, Anticorrosive pigment, electrochemical test, Nubirox

P-42 Identifying the individual contributions of T_g and reversible interactions in near T_g healing polymers: a high resolution rheology-based deconvolution protocol.

Vincenzo Montano, Sybrand van der Zwaag, Santiago J. Garcia
Novel Aerospace Materials, Aerospace Engineering Faculty, Delft University of Technology, Delft, the Netherlands

Polymers and coatings healing at temperatures close to their T_g (e.g. RT healing polymers) make it difficult to understand the healing governing process and the individual role of embedded reversible interactions. This is due to relaxation phenomena associated to reversible interactions and segmental mobility (T_g) overlapping in the frequency domain. The development of tools allowing the deconvolution into the individual contributions will highly benefit the development of novel SH polymers for coating applications. We address this challenge in the current work.
One example of such healing polymers with overlapping relaxation phenomena was observed for the room temperature healing of brushed polymers where healing was attributed to the interactions among aliphatic branches (RT SH-PEIs)[1].
In order to be able to identify the contribution of relaxation phenomena taking place in the same frequency domain we explore the use of tools capable of identifying the energetic contribution of each individual component.
For the purpose a high resolution rheology based approach capable of discerning the unique role of different reversible interactions in the overall relaxation process of self-healing systems is proposed (Fig. 1). The approach is based on the fitting of the experimentally calculated relaxation spectra to the well-known generalized Maxwell (or Maxwell-Wiechert, Eq. 1) model. The approach allows the deconvolution of broad relaxation phenomena into narrower relaxation peaks. These can be associated to specific macromolecular relaxation phenomena following detailed analysis.

HMaxwell-Weichert(τ) = ∑ni=1 Ai exp(-(ln(τ)-ln(τi))2/2σi2) Eq. 1

The validity of the approach was verified on the mentioned RT SH-PEIs 1, an ad-hoc designed self-healing branched polyamide and non-branched non-healing polymers for which just segmental mobility was expected.
The protocol allows to get a better insight of the individual contributions to macroscopic healing. The computed model constants add quantitative information about the energy dissipated and the time shift of the specific relaxation transitions allowing for a better correlation between polymer architecture and macroscopic healing (i.e. effect of branches content and effect of backbone rigidity).
Based on the obtained results so far, it is expected that this protocol will help improving our understanding of the contribution of individual frequency-overlapping phenomena on macroscopic polymer behaviour.

References:

1.Susa, A., Bose, R. K., Grande, A. M., Van Der Zwaag, S. & Garcia, S. J. Effect of the Dianhydride/Branched Diamine Ratio on the Architecture and Room Temperature Healing Behavior of Polyetherimides. ACS Appl. Mater. Interfaces 8, 34068–34079 (2016).

Keywords: Branched Polymers, Rheology, Self-healing

P-43 Electrodeposition of Zn from $ZnCl_2$-ethyl methyl sulfone electrolyte

Essam Moustafa, A. Dietz, M. Thomas
Fraunhofer IST, Braunschweig, Germany

The electrodeposition of Zn on to steel substrates from $ZnCl_2$ - ethyl methyl sulfone (EMS) eutectic mixture was studied. The electrochemical behaviour of electrolyte was investigated using cyclic voltammetry (CV). The electrodeposition process is highly reversible and the cathodic and anodic currents increase with higher temperatures. The surface morphology, purity and crystallinity of as deposited Zn were assessed by scanning electron microscopy (SEM), energy dispersive X-ray analysis (EDAX) and X-ray diffraction (XRD), respectively. At room temperature, discrete coarse hexagonal-shaped Zn particles were formed; however, the grain size and thickness increase with increasing temperature because of increasing of mobility of the electroactive species towards the cathode surface. The results of X-ray diffraction (XRD) and energy dispersive X-ray analysis (EDAX) confirmed that the as-deposited layer is crystalline Zn with high purity.

P-44 Evaluating the effect of graphene dispersion on thermal conductivity of waterborne acrylic coats

Babak Nazari Loskokelayeh, Zahra Ranjbar
Department of Surface Coating and Novel Technology, Institute for Color Science & Technology, Tehran, Iran

Usage of thermal conductive coatings has become increasingly significant due to numerous applications in various industries. According to inappropriate thermal conductivity of polymeric matrixes, conductive nanofillers are used in coatings compositions more and more. Graphene as an extraordinary filler with outstanding thermal properties has been considered more than other substances. The inimitable thermal properties of graphene have made it possible to enhance thermal conductivity of polymer composites. In this study, the effects of different amounts of graphene nanoplatelets (GNPs) on thermal conductivity of waterborne acrylic coats have been demonstrated. Due to the hydrophobic nature of GNPs and intermolecular forces among them, stable dispersion of graphene in water can be achieved utilizing surfactants. The character of surfactants (ionic and non-ionic) led us investigate the stability mechanisms in aqueous dispersions. Zeta potential measurement was done for samples containing ionic surfactant, which data were evaluated through DLVO theory. By evaluating the performed experiments and solving the stability equations of DLVO theory, Cetyltrimethylammonium Bromide (CTAB) as a cationic surfactant shown the best performance in dispersion and stability of GNPs in aqueous media. Waterborne acrylic coatings with different percentage of GNPs were prepared by adding stable dispersion of graphene in polymeric matrix. Optical characterization was carried out to identify how GNPs spread into polymeric phase. In order to measure the specific heat capacity of samples, Differential Scanning Calorimetry (DSC) was done. By using laser beam, thermal diffusivity of coatings in terms of weight fraction of GNPs has been measured. Calculation of thermal conductivity was carried out by the values obtained from diffusivity and specific heat capacity. The results indicate that by increasing the amount of GNPs to 1wt. %, thermal conductivity of coating has reached to 1.661 W/mK while thermal conductivity of pure acrylic was 0.76 W/mK. Enhancement of thermal conductivity can be attributed to the formation of a suitable network among GNPs inside the coating. Therefore, as the weight percentage of fillers increases, empty spaces among them is relieved and more favorable heat path for transmission and heat release is provided as a result of lower thermal strength.

References:

[1] A.K. Geim, K.S. Novoselov, The rise of graphene., Nat. Mater. 6 (2007) 183–91. doi:10.1038/nmat1849.
[2] A.A. Balandin, S. Ghosh, W. Bao, I. Calizo, D. Teweldebrhan, F. Miao, C.N. Lau, Superior thermal conductivity of single-layer graphene., Nano Lett. 8 (2008) 902–7. doi:10.1021/nl0731872.
[3] R.R. Nair, P. Blake, A.N. Grigorenko, K.S. Novoselov, T.J. Booth, T. Stauber, N.M.R. Peres, A.Geim, Fine structure constant defines visual transparency of graphene., Science. 320 (2008) 1308. doi:10.1126/science.1156965.
[4] J.S. Bunch, S.S. Verbridge, J.S. Alden, A.M. van der Zande, J.M. Parpia, H.G. Craighead, P.L. McEuen, Impermeable atomic membranes from graphene sheets., Nano Lett. 8 (2008) 2458–62. doi:10.1021/nl801457b.
[5] M. Fang, K. Wang, H.B. Lu, Y. Yang, and S. Nutt, "Single-layer Graphene Nanosheets with Controlled Grafting of Polymer Chains", Mater. Chem., 19, 2009, 7098– 7105.
[6] J.Y Choi, S.W. Kim, and K.Y. Cho, " Improved Thermal Conductivity of Graphene Encapsulated Poly(methyl methacrylate) Nanocomposite Adhesives with low Loading Amount of Graphene", Compos. Sci. Technol., 2014, 147–154.
[7] F. Yavari, H. Raeisi Fard, K. Pashayi, Zh. Yu, T. Borca-Tasciuc, N. Koratkar, Enhanced Thermal Conductivity in a Nanostructured Phase Change Composite due to Low Concentration Graphene Additives, J. Phys. Chem. C 2011, 115, 8753–8758. doi: 10.1021/jp200838s.
[8] Y-X. Fu, Zhuo-Xian He, Dong-Chuan Mo, Shu-Shen Lu, "Thermal conductivity enhancement of epoxy adhesive using graphene sheets as additives", 2014, 276-283.

Keywords: Graphene Dispersion, Thermal Conductivity, Waterborne Conductive Coatings

P-45 Triphenylamine based hyperbranched polyimides for coating technologies: 1. Synthesis and characterization

Naila Khalid[1], Humaira Siddiqi [2], Asma Iqbal[2], O Ok Park[3]
[1]Chemistry, Kohat University of Science and Technology, KUST, Kohat, Pakistan
[2]Chemistry, Quaid-i-Azam University, Islamabad, Pakistan,
[3]Department of Chemical and Biomolecular Engineering, Korean Avanced Institute of Science an Technology, Kaist, Daejeon, South Korea

A trifunctional monomer triamine N^1-(4-(4-aminophenoxy)phenyl)-N^1-(4-aminophenyl)benzene-1,4-diamine (TA) was successfully synthesized by cesium fluoride mediated nucleophilic N-arylation, followed by catalytic reduction. The polymerization of monomer triamine (TA) with a series of aromatic dianhydride was carried out by two-step polycondensation process. The dianhydride monomers: 1,2,4,5-tetracarboxylic dianhydride (PMDA), 2,2'-bis(3,4-dicarboxyphenyl)hexafluoropropane dianhydride (FDA), 3,3′,4,4′-benzophenone tetracarboxylic dianhydride (BTDA) and 4,4'-oxydiphthalic dianhydride (OPDA) were employed. Thus a series of essentially colorless aromatic tree-shaped hyperbranched polyimides (TA-PIs) spin-coated thin films of type A_2B_3 were fruitfully synthesized. The apposite molar ratios of TA to a dianhydride employed were 2:3 for TA-PIs synthesis using A_2B_3 approach. The synthesized HBPIs exhibited moderate number-average molecular weights (25666-28367 g/mol) and polydispersity in the range 2.6-3.0, as revealed by gel permeation chromatography (GPC) measurement. The optical transparency of spin-coated polymeric films was evaluated by UV-vis spectral analysis, indicating UV-absorption below 400 nm in the range 256-268 nm. The optical transparency is an essential requirement for their application as second surface mirror coating of thermal control systems. Thermal degradation analysis of the TA-PIs showed that synthesized materials are thermally stable up to 450 ^{o}C, and DSC measurement indicated T_g in the range 290-302 ^{o}C.

P-46 Pre-treatment study of varied woods with high performance light stabilizers

Yung-Chi Yang[1, 2]
[1]Everlight, Taipei, Taiwan, Republic of China,
[2]Triconor Distribution B.V. Soest, the Netherlands

Lignin, when exposed to sunlight can easily be degraded. To preserve the beauty of wood grain, UV absorber and light stabilizer containing waterborne clear coats are commonly applied to the surface for protection. However, the effectiveness of such application is limited. According to our studies, pre-treatment of varied wood surface prior to the application of clear top coat can effectively extend the protection of lignin from UV light degradation. Eversorb AQ-1, a blend of UV absorber and HALS, is developed by Everlight Chemical Industrial Corporation (ECIC) specifically for waterborne clear coats. Eversorb SB-1, designed for wood pretreatments, can protect lignin from UV light degradation. This paper is a comparison study in accelerated weathering of varied wood with and without SB-1 pre-treatment in different co-solvents. Design of Experiment (DOE) was used in this study. Our data confirmed that pre-treatment is better than not pre-treatment from UV light degradation. Lignin is a complex organic polymers. The funtional group contains hydroxyl groups (OH). Two types of co-solvents are tested in this system, the solubility parameter of isopropyl alcohol is 11.5 and butyl carbitol is 9.9. After weathering test, SB-1 mixed with isopropyl alcohol is better than butyl carbitol. Two softwoods contain 25-35% lignin and hardwoods contain 18-25% lignin. Theory, SB-1 pre-treatment is significant to protect softwoods. Data indicates that pre-treatment with SB-1 in softwoods (Southern Pine and Douglas fir) are significant effect than hardwoods (Cedar, Southern beech and African Teak)

Keywords: Additive, UV protection, Waterborne coating, wood coating

P-47 Effect of acrylic binder type and calcium carbonate amount on paint properties

Verónica Alvarez [1], Maria Paulis[1, 2]
[1]Quimica Aplicada, San Sebastian, Spain
[2]Polymat, San Sebastian, Spain

The final price of waterborne paints can be reduced by the addition of fillers without significant losses of the films properties up to a certain amount of filler addition. Therefore, it is important to study the interaction between binders and fillers, which is the objective of this work. To accomplish the objective of this work, different filler/polymer latex blends have been prepared using as less additives as possible and the effects of the latex nature (in terms of surface coverage and type and amount of functional monomer employed during the synthesis) on the film properties (scrub resistance, tensile properties, water uptake and gloss) have been investigated. Calcium carbonate ($CaCO_3$) and poly(methyl methacrylate-co-butyl acrylate) P(MMA/BA) have been used as extender and latex respectively.
It was found that better final paint performances were obtained (better mechanical and scrub properties and less water absorption) when paints were prepared using latexes synthesized at acidic conditions in comparison with paints prepared using latexes synthesized under basic conditions. Moreover, in order to obtain stable latex/$CaCO_3$ blends, the use of a functional monomer during the synthesis of the latex was necessary. Regarding the functional monomers employed during the synthesis of the latex, both acrylic acid (AA) and methacrylic acid (MAA), 1 weigh based on monomer percent (wbm%), were used. Paints prepared with the latexes synthesized with MAA functional monomer presented better final paint performances (less water uptake, and higher scrub and mechanical resistance). Increasing the amount of the functional monomer from 1 wbm % to 3 and 5 wbm %, final paint properties changed (final paints absorbed more water, had worse scrub resistance and more rigid mechanical properties).

The authors would like to thank the financial support provided by the Industrial Liasion Program of Polymat, the Basque Country Government (IT999-16), and the University of the Basque Country UPV/EHU (UF11/56)

Keywords: Acrylic latex, Calcium Carbonate filler, Film properties

P-48 Utilizing electrochemical impedance spectroscopy and simultaneous kelvin probe-Raman spectroscopy to investigate the diffusion/migration of electrolyte along the metal/water-borne coating interface

Shima Khazraee Sheidaee Mehr[1], Michael Dornbusch[1], Wolfgang Bremser[2]
[1]Hochschule Niederrhein, Krefeld, Germany
[2]Paderborn University, Paderborn, Germany

Water-borne organic coatings are eco-friendly coatings extremely applied to protect metallic substrates from corrosion. In this work, several water-borne coatings were prepared and studied precisely to achieve the best model polymer coating toward corrosion protection. The barriers properties of the coating towards water diffusion is so important because the presence of the water causes loss of adhesion and acts as the medium for the diffusion of oxygen and aggressive ions. Therefore, the capacity of the coating which is an important parameter and displays the water up-take in the coating measured by electrochemical impedance spectroscopy (EIS). Furthermore, the mechanisms behind diffusion or migration of corrosive species at the metal-polymer interface plays an important role in the long-term performances of the coated materials. Hence, to understand the fundamental processes of the electrolyte transport at metal-polymer coating and the delamination underneath of the model polymer coatings on steel panels the powerful technique such as simultaneous Kelvin Probe(KP) with Raman spectroscopy was utilized. Furthermore, microscopic methods such as SEM/EDX or AFM were applied to observe the microstructure of the coating and the plasticizing behaviour of the polymer. Simultaneously, the technical clear and pigmented organic coating were characterized using spectroscopic method such as infrared spectroscopy in attenuated reflection (IR-ATR), Raman and UV-VIS to analysis the effect of rust inhibitors and polymer structure in reducing of cathodic dis-bonding of a coating and delamination rate.

References:

[1] R. Posner, A. M. Jubb, G. S. Frankel, M. Stratmann, H. C. Allen, A simultaneous Kelvin Probe and Raman spectroscopy approach for in situ surface and interface analysis, Electrochim. Acta 76 (2012) 34.

Keywords: diffusion/migration of electrolyte , electrochemical impedance spectroscopy , Kelvin Probe(KP), Water-borne organic coatings

P-49 Advanced corrosion techniques to characterized waterborne primers performance.

Sonny Ngo [1], [2]
[1]MERI, Sheffield Hallam University, Sheffield, United Kingdom
[2]Long Term Development Lab, Becker Industrial Coatings, Liverpool, United Kingdom

Coil coating polyester topcoat were coated over four different primers. The systems were investigated in natural and accelerated environments. Their performances were characterised by electrochemical impedance spectroscopy (EIS) and focused ion beam/ scanning electron microscope (FIB/SEM) to determine the extent and mode of corrosion failures. Standard coil coating primer performed excellently in natural and accelerated environments. However, non-pigmented waterborne primer performed poorly, as you would expect for a coating containing no pigmentation. The 2 waterborne primers containing anticorrosive pigments with and without conversion technology show an improvement in the corrosion protection of hot dipped galvanised steel. EIS and FIB/SEM analysis strongly supported these findings.

AUTHOR INDEX

Abdelrahman, A.	S02-1
Abohalkuma, T.	S15-3
Adan, O.	S24-3
Afanaseva, D.	P-24
Aktas, L.	S10-2
Albers, P.	P-15
Al-Hellani, R.	S32-3
Ali, B.	S16-3
Alibakhshi, E.	P-25
Allonas,X.	S04-1
Al-Masri, M.	S14-3
Alvarez, V.	P-47
Amini, M.	P-27
Amini, M.	S06-2
Arabi, A. M.	P-12
Arturo, V.-Z.	P-39
Asadi, N.	P-26
Ascencio, M.	S23-1
Asua, J. M.	S16-2,S32-2
Aydogan, O.	S27-1,S27-2
Ballard, N.	S16-2
Banzduch, J.	P-22
Bauer, N.	S18-3
Bayminov, B.	P-28
Becker-Willinger, C.	S18-3,S16-3
Bedford, O.	S30-2
Behgam, R.	P-03
Bernhofer, T.	S17-3
Bertin, E.	S29-3
Biuk Afshari, B.	P-04,P-06
Bohacik, P.	P-22
Bohorquez, S.	S13-1,S32-2
Bongiovanni, R.	S20-1
Bos, M.	S03-2
Bosma, M.	S22-3
Bourebrab, M.	S21-2
Boye, A.	P-29
Bramlage, C.	S19-2
Brannstrom, S.	S12-1
Breivogel, A.	P-02
Bremser, W.	P-48
Brett, C.	S25-2,S35-1,S35-2
Briand, A.	S29-3
Broeders, J.	S01-3
Broer, D. J.	S08-1
Brück, S.	S12-2
Brünink, D.	S19-2
Brunke, J.	S16-3,S18-3
Cairns, C.	S33-3
Chen, Q.	S08-3
Christ, U.	S02-2
Cordeiro de Castro, C.	S15-2
Croll, S.	S28-3
Croutxe-Barghorn, C.	S04-1,S23-3
Crowston, E.	S07-2
Curtis, A.	S29-3
Daamen, R.	S26-3
Dam-Johansen, K.	P-33,P-29,P-35
Dargatz, M.	S32-3
Davies, T.	S28-2
De Backer, S.	P-30
De La Cal, J. C.	S32-2
De Oliveira, P. W.	S16-3
De Sousa Gomes, L.	S29-1
De With, G.	P-15
Debinski, P.	P-29
Denisova, N.	P-24
Denissen, P. J.	S10-1
Diebold, M.	S30-1
Dietliker, K.	S04-2
Dietz, A.	P-43
Dollani, H.	S29-3
Domnick, J.	S19-3
Dornbusch, M.	P-37,P-07,P-48
Durand, G.	S21-2
Ebrahimi, M.	S18-2, P-05
Ehlers, J.-E.	P-37
Eichenberger, F.	S04-3,S17-2
Entenmann, M.	S31-2
Erich, B.	S24-3
Esteves, A.	P-15
Falkenberg, H. B.	P-33
Fašung, D.	S33-2
Faupel, F.	S35-2
Féat, A.	P-14
Federle, W.	P-14
Felhosi, I.	P-13
Finnveden, M.	S12-1
Florio, J.	S07-3
Fricke, B.	S35-1
Fritzen, P.	S29-3
Gallardo-Hernández, E. A.	P-39
Galvan-Miyoshi, J.	S32-1
Ganjaee, M.	P-06
Garcia , S. J.	S10-1,S16-1,P-42
García-Valenzuela, A.	S32-1
Gariepy, K.	S13-3
Garray, D.	S31-1
Gedan-Smolka, M.	S31-2,S31-1
Ghasemi, E.	P-25
Gillet, S.	P-34
Gioia, C.	S05-2
Gorji, S.	P-40
Göttl, N.	S02-2
Govers, S.	P-15
Gravendeel, M.	S36-3

Greisiger, H. S31-2
Grianko, I. P-23
Groenewoud, M. S07-3
Gueller, R. S22-1
Gulbahar, B. S27-1,S27-2
Günkaya, G. P-19
Gustini, L. S12-3
Gutmann, J. P-07,P-37
Haddadi, S. P-27
Hammer, P. S02-3
Hantson, A.-L. S15-2
Harb, S. S02-3
Heinonen, S. S14-1
Herrwerth, S. S33-1
Herszenhaut, M. S11-1
Heuer, M.-Y. S04-3,P-31
Heydari, G. S14-1
Heymans, D. S11-1
Hietbrink, R. S33-3
Hijnen, N. S33-3
Hilt, M. S10-2
Hiltl, S. S30-2
Hinz, A. S35-2
Hoffman, L. S20-3
Hrebicik, M. S07-2
Huang, Y. P-21
Huang, A. S20-3
Huinink, H. S24-3
Ibrahim, A. S04-1
Iliev, B. S24-2
Ilin, A. P-23
Imbesi, P. S03-2, S07-1
Inauen, J. S15-1
Indeikin, E. P-23,P-24
Iqbal, A. S23-2,P-45
Ismael, R. S30-3
Jacob, H.-J. P-32
Jamshidi, M. P-04
Jana, S. S34-3
Janssen, W. S12-3
Jawerth, M. S05-2
Jiang, L.-M. S23-1
Johansen, K. D. P-11
Johansson, M. S12-1,S05-2, P-01,
Jonathan, Z.-G. P-39
Jongerius, M. S33-3
Jørgensen, S. B. P-33
Jung, H.-J. P-20
Kadlecova, M. P-22
Kainz, Q. S32-3
Kaleva, A. S14-1
Kali, G. S18-3
Kamperman, M. P-14
Keddie, J. PL-02
Kern, N. S19-1
Keshmiri, N. P-40
Keuerleber, M. S22-2
Khaki, I. P-04,P-06
Khalaj Asadi, A. S18-2
Khalid, N. S23-2,P-45
Khatai, L. S13-3
Khazraee Sheidaee Mehr, S. P-48
Kiil, S. P-33,P-11,P-29,P-35,P-21
Kilpeläinen, V. S21-3
Kittler, T. S13-3
Klein, S. E. S12-2
Kleinsteinberg, F. S19-2
Komorowski, L. P-08
Koning, C. S12-3
Kooistra, F. S31-3
Kopec, M. S28-2
Körstgens , V. S35-2
Kosolapov, A. P-28
Kostitsyn, A. P-23
Krawczyk, T. P-07
Kreisig, S. S08-3
Królikowska, A. P-08
Küchenmeister-Lehrheuer, C. S36-3
Kuczyńska, H. P-08
Kühnert, I. S31-1
Kuyucu, O.G. S27-1,S27-2
La Rosa, A. S21-2
Laarmann, T. S35-1
LabbaniMotlagh, A. P-25
Lahem, D. S15-2
Lange, C. S14-1
Langer, E. P-08
Lavilla, C. S12-3
Lawoko, M. S05-2
Legrix-Elton, A. S29-3
Leonardo, I.F.-C. P-39
Leunissen, R. S09-2
Levänen, E. S14-1
Li, K. P-09
Li, Q. P-17
Li, S. P-17
Lin, H.-W. P-18
Lista, G. S01-2
Liu, B. P-10
Loh, F. S33-3
Löhrer, F. C. S35-2
Loos, J. S08-3
Lopez, A.B. S32-2
Lucas, F. S32-3
Ludewig, M. S11-3
Lukacs, A. P-13
Lundmark, S. S05-2
Lunzer, F. S13-3
Luo, S. P-11
Lutke Schipholt, N. S26-3
Lyon, S. S28-2
Ma, G. P-17
Machotova, J. P-36
Mahdavian, M. S06-2, P-12,P-26,

P-03,P-25,P-27
Mahmoud, H. S18-1
Maillard, C. S29-3
Malmström, E. S12-1
Manea, M. S11-2
Manni, J. S14-1
Manuel, V.-T. P-39
Marcos, C. S22-1
Marisa, M.-R. P-39
Markkula, A. S14-1
Martinelle, M. S12-1, P-01
Martinez de Ilarduya, A. S12-3
Meeuwisse, M. S32-2
Meier, I. S17-2
Memisoglu, G. S27-1,S27-2
Mertens, T. S10-3
Mestach, D. S13-1,S32-2
Michael, G. P-31
Ming, W. M. S36-1
Miotto, C. S03-2, S07-1
Miri, B. P-05
Mittal, N. S25-2
Mohamadzade Attar, M. R. P-05
Mohammadkhani, R. P-38
Mohseni, M. S18-2,P-40
Molhoek, L. S03-2
Molnar-Nagy, L. P-13
Montano, V. P-42
Mousavi, M. P-12
Moustafa, E. P-43
Müller, N. S16-3
Müller-Buschbaum, P. S35-2
Münch, A. S20-2
Muñoz-Guerra, S. S12-3
Na, M.-J. P-20
Naderi, R. S06-2, P-26
Najmi, P. P-40
NameerS. P-01
Nazari Loskokelayeh, B. P-44
Nevejans, S. S16-2
Ngiabprasert, B. S11-1
Ngo, S. P-49
Nikkanen, J.-P. S14-1
Ninnemann, F. S03-3
Nothhelfer-Richter, R. S02-2
Ohm, W. S25-2,S35-1
Okey-Onyesolu, F. S09-1
Okumuş, Y. P-19
Oldörp, K. S36-3
Olivares, A. S32-1
Olsen, S. M. P-21
Omanovic, S. S23-1
Omeis, J. PL-03
Oro, E. P-41
Panasiuk, M. S29-2
Park, S.-D. P-20
Park, O. O. S23-2,P-45
Paulis, M. P-47
Paunikallio, T. S14-1
Perez, P. P-41
Peters, R. S08-3
Pilger, J. S17-3
Pirhady Tavandashti, N. P-38
Pitarch, J. S36-3
Plebst, S. S24-2
Podzimek, S. P-22,P-36
Polonskyi, O. S35-2
Popova, N. P-28
Pozman, T. P-13
Price, C. S33-3
Prosek, T. S06-1
Qu, J.-E. S23-1
Rafiei Hashjin, R. S25-1
Ramazani, S.A. A. P-27
Ramezanzadeh, B. S35-3,P-12,P-25
Ramezanzadeh, M. S35-3
Ranjbar, Z. S25-1,P-44
Rannee, A. S23-3
Ravichandran, R. S07-3
Reck, B. S16-2
Rehahn, M. S12-2
Retailleau, M. S04-1
Reuter, E. S19-2
Reynolds, K. S33-3
Ring, L. P-35
Roland, K. S33-1
Roller, S. S32-3
Romanowska, C. S11-1
Rommens, J. S30-1
Rossenaar, B. S28-2
Rostami, M. P-04,P-06
Roth, S. V. S25-2,S35-1, S35-2
Rozek, L. S34-2
Rtimi, S. S25-3
Rueger, J. S36-3
Saarimaa, V. S14-1
Sanaei, Z. S35-3
Sanchez, T. S26-1,P-34
Santilli, C. S02-3
Sapozhnikov, D. P-28
Schaller, C. S06-3
Scheerder, J. S24-3
Schmitz, C. S27-3
Scholz, P. S31-1
Schrader, L.-U. S21-1
Schreiner, C. S21-1
Schubert, T. S24-2
Schulze, M. S12-2
Schwartzkopf, M. S35-1,S35-2
Semjonov, S. P-28
Semlitsch, S. P-01
Senechal, T. S15-2
Shaban, A. S15-3
Sharif, F. P-05

Shen, B. S19-3
Shkirskiy, V. S26-1
Shojaei, A. P-38
Siddiqi, H. S23-2,P-45
Siegmann, K. S15-1
Skopintseva, N. P-24
Söderberg, D. S25-2,S35-1
Soloveva, V. P-23
Spacek, V. P-22
Sprakel, J. S08-2
Sproncken, C. S36-3
Steiner, A. S13-3
Stenzel, V. S21-1
Sterchi, R. S15-1
Sterckx, L. S13-3
Strehmel, B. S27-3
Struck, S. S17-2
Struck, O. S30-2
Strunskus, T. S35-2
Stuettgen, A. S04-3
Suwala, K. S29-2
Synytska, A. S24-1
Szewczyk, J. S03-2, S07-1
Takeno, S. S14-2
Taylor, P. P-14
Taylor, A. S21-2
Tejada, E. S11-3
Telegdi, J. S15-3
Terryn, H. S28-1
Thierry, D. S06-1
Thomas, M. P-43
Thyssen-Wallner, V. S14-3
Tian, M. S08-3
Tiedje, O. S19-3
Timothy, U. S09-1
Tomozeiu, N. S17-1
Trentin, A. S02-3
Tromsdorf, U. S32-3
Tsujita, J. S14-2
Ty, C. S06-3
Uhlmann, P. S20-2
Uzoh, C. S09-1,S05-3
Väisänen, P. S14-1
Valka, R. P-36
Van Benthem, R. P-15
Van De Kamp-Peeters, L. S24-3
Van de Velde, B. S18-1
Van der Gucht, J. P-14
Van der Meulen, J. S01-1
Van der Slot, S. S08-3
Van der Ven, L. P-15
Van der Zalm, M. S33-3
Van der Zwaag, S. P-42
Van Haveren, J. S05-1
Van Leerdam, K. S28-2
Van Loon, S. S03-1,S21-3
Van Oorschot, A. S09-2
Vega, B. S11-3
Verhaege, A. S30-1
Verstuyft, L. S29-3
Villamaina, D. S15-1
Vitale, A. S20-1
Vivier, V. S26-1
Volovitch, P. S26-1,P-34
Voogt, B. S24-3
Vygodskii, Y. P-28
Vyörykkä, J. S03-2, S07-1
Wang, T. P-11
Wang, X. P-21
Wanner, M. S10-2
Weber, K. S22-2
Webster, D. S36-2
Wegmann, V. S22-2
Weinell, C. E. P-11,P-29,P-21,P-33,P-35
Wekking, J. S09-3
Wenz, G. S18-3,S16-3
Widjaja, W. S33-3
Wijnen, M. S33-3
Willems, N. S03-2
Winkler, M. S15-1
Witzleben, S. S12-2
Wörheide, R. J. S26-2
Wysoglad, J. P-37
Yang, H. P-20
Yang, Y.-C. P-46
Yari, H. S25-1
Ye, Q. S19-3
Yvon, E. S34-1
Zaldo, F. S32-1
Zeng, Y. P-35
Zielecka, M. S29-2
Zubielewicz, M. P-08

ETCC 2018
https://etcc2018.org

ISBN 978-1-5108-7959-1

CURRAN ASSOCIATES INC.
proceedings.com

Systems Security, Resilience, and Emergency Management 2018

Transportation Research Record: Journal of the Transportation Research Board

Volume 2672, Issue 1

SAGE Publishing